Level B

Investigating
Diversity and Limits

Middle School Science & Technology

Innovative Science Education
founded 1958

KENDALL/HUNT PUBLISHING COMPANY
Dubuque, Iowa

0-8403-6678-7

Copyright © 1994 by the BSCS. All rights reserved. No part of this work may be reproduced or transmitted in any form or by any means, electronic or mechanical, including photocopying and recording, or by any information storage or retrieval system, without permission in writing. For permissions and other rights under this copyright, please contact the BSCS, 830 North Tejon, Suite 405, Colorado Springs, Colorado 80903.

This material is based on work supported by the National Science Foundation under Grant No. MDR 8855657. Any opinions, findings, conclusions, or recommendations expressed in this publication are those of the authors and do not necessarily reflect the views of the granting agency.

BSCS Development Team

Rodger W. Bybee, *Principal Investigator* (1988–92)
Janet Carlson Powell, *Project Director* (1988–92)
Kathrine A. Backe, *Staff Associate, Implementation* (1991–92)
Wilbur C. Bergquist, *Staff Associate, Evaluation* (1991–92)
Deirdre Binkley-Jones, *Project Secretary* (1992)
Jan Chatlain Girard, *Art Coordinator* (1989–92)
Sariya Jarasviroj, *Production Assistant* (1992)
Terri Johnston, *Project Secretary* (1991–92)
Donald E. Maxwell, *Staff Associate, Staff Development* (1990–92)
Mary E. McMillan, *Staff Associate, Curriculum Development* (1991–92)
Josina Romero-O'Connell, *Staff Associate, Curriculum Development* (1991–92)
Teresa Powell, *Project Secretary* (1989–92)
Judith Martin Rhode, *Research Assistant*, (1989–91), *Staff Associate, Curriculum Development* (1992)
Joe Ramsey, *Production Assistant* (1992)
William C. Robertson, *Staff Associate, Curriculum Development* (1989–91)
Nancy Smalls, *Project Secretary, Graphics* (1990–92)
Jenny Stricker, *Staff Associate, Curriculum Development* (1992)
Pamela Van Scotter, *Staff Associate, Editing* (1990–92)
Lee B. Welsh, *Production Coordinator* (1989–92)
Yvonne Wise, *Project Secretary* (1989–92)

These titles and dates indicate the primary area of responsibility for each person and the years they worked on the project. Everyone on the project team contributed in numerous ways to create this curriculum.

Photography

Carlye Calvin
NASA
Unicorn Stock Photos—Robert W. Ginn
Biosphere II—Scott McMullen
Richard B. Levine
NAMES Project—Marcel Mirand
Shawn Sigstedt
Unicorn Photos—Joseph L. Fontenct
Dr. Martin Lockley
Charles W. Melton
R. E. Barber
Colorado Springs Utilities
G. M. Hughs Electric
Vanuga Photography
Frances M. Roberts
Mark Schug
Les Van
Student photo on cover—John W. Clark Photography

Artists for First Commercial Edition

Jan Chatlain Girard
Linn and Bob Trochim—Animart
Susan Bartel
Nancy C. Smalls
Janet Huntington-Hammond
Carmen Franco-Stephenson
Bill Ogden—Animation Renderings
PC&F, Inc.—Technical Illustrations

BSCS Administrative Staff

Roger G. Olstad, *Chair, Board of Directors*
Joseph D. McInerney, *Director*
Rodger W. Bybee, *Associate Director*
Larry Satkowiak, *Chief Business Officer*

Board Members

Elliot Asp, *Littleton Public Schools, Littleton, Colorado*
Randall Backe, *Kansas State University, Manhattan, Kansas*
Pat Barry, *Wilbur Wright Middle School, Milwaukee, Wisconsin*
Bonnie Brunkhorst, *California State University, San Bernardino, California*
Herbert Brunkhorst, *California State University, San Bernardino, California*
H. Mack Clark, *Air Academy District #20, Colorado Springs, Colorado*
Mary Doyen, *Rocky Mountain Center for Health Promotion and Education, Northglenn, Colorado*
Linda Ganatta, *Timberview Middle School, Colorado Springs, Colorado*
April Gardner, *University of Northern Colorado, Greeley, Colorado*
Cynthia Geer, *University of Cincinnati, Cincinnati, Ohio*
Merton Glass, *University of South Florida, Tampa, Florida*
Johnnie P. Hamilton, *Franklin Intermediate School, Chantilly, Virginia*
Debbie Hill, *Eagleview Middle School, Colorado Springs, Colorado*
David Housel, *Oakland Schools, Waterford, Michigan*
Roger Hubley, *Pleasant Run Middle School, Cincinnati, Ohio*
Paul DeHart Hurd, *Professor Emeritus, Stanford University, Palo Alto, California*
Candace Julyan, *Technical Education Research Centers, Cambridge, Massachusetts*
David Kennedy, *State Department of Education, Olympia, Washington*
Joyce Kerce, *W. D. Sugg Middle School, Bradenton, Florida*
Keith Kester, *Colorado College, Colorado Springs, Colorado*
Julie Kropf, *Hollenbeck Middle School, Los Angeles, California*
Thomas Liao, *SUNY, Stony Brook, New York*
Thomas Lord, *Indiana University of Pennsylvania, Indiana, Pennsylvania*
Susan Loucks-Horsley, *The NETWORK, Andover, Massachusetts*
Glenn Markle, *University of Cincinnati, Cincinnati, Ohio*
James McClurg, *University of Wyoming, Laramie, Wyoming*
Francesca Mollura, *Academy of Liberal Arts and Sciences, Kansas City, Missouri*
Cathy Oates, *Challenger Middle School, Colorado Springs, Colorado*
Michael Padilla, *University of Georgia, Athens, Georgia*
Rita Patel-Eng, *SUNY, Stony Brook, New York*
E. Joseph Piel, *Professor Emeritus, SUNY, Stony Brook, New York*

(Continued on p. 333).

Table of Contents

Preface	ix
Program Overview	xi

UNIT 1 — What Is Normal? — 1
COOPERATIVE LEARNING OVERVIEW — 2

CHAPTER 1 — Identifying Limits and Diversity — 5

Engage/Explore	**INVESTIGATION:** Star Tracers	6
Explore	**INVESTIGATION:** Threading the Needle	9
Explore	**CONNECTIONS:** How Do You Spell Success?	11
Explore	**INVESTIGATION:** If at First You Don't Succeed	11
Explain	**READING:** Doing It All the Same	13
Elaborate	**INVESTIGATION:** Seeing the World Around You	15
Elaborate	**READING:** The Human Eye and Peripheral Vision	19

CHAPTER 2 — Normal Ranges of Limits and Diversity — 23

Engage/Explore	**INVESTIGATION:** The Individuality of Popcorn	24
Explain	**READING:** What Is Normal?	26
Elaborate	**INVESTIGATION:** A Diversity of Popcorn	29
Elaborate	**CONNECTIONS:** More on the Meaning of the Normal Curve	30
Explain	**READING:** The Value of the Normal Curve	31
Elaborate	**INVESTIGATION:** The Normal Curve and You	33
Elaborate	**INVESTIGATION:** A Flag of a Different Color	35
Elaborate	**READING:** The Diversity of Afterimage	37

iii

CHAPTER 3

Using Limits to Set Standards — 41

Engage	**INVESTIGATION:** What Do You Really Know about TV?	42
Explore	**INVESTIGATION:** Taking a Closer Look at TV Pictures	43
Explore	**CONNECTIONS:** The Ultimate TV	45
Explore	**READING:** TV Pictures and Color TV	45
Explore	**INVESTIGATION:** A Learning Journey	48
Explore	**STATION 1:** Constructing a Spinner	49
Explore	**STATION 2:** Continuous Motion through Animation	50
Explore	**STATION 3:** Continuous Motion at the Movies	52
Explore	**STATION 4:** Flicker-Fusion Frequency	54
Explore	**STATION 5:** How Many Lines Per TV Picture?	56
Explore	**STATION 6:** *TV Teasers* Computer Station	57
Explore	**WRAP UP FOR ALL STATIONS**	58
Explore	**CONNECTIONS:** Your Experiences at the Stations	58
Explore	**INVESTIGATION:** The Optimal TV Viewing Distance	59
Explain	**READING:** Setting Standards and Human Factors	60
Elaborate	**CONNECTIONS:** Clear As a Belle	61
Elaborate	**READING:** The National Television Systems Committee Uses Human Factors	63

CHAPTER 4

Using Diversity to Set Standards — 67

Engage/Explore	**INVESTIGATION:** Watching *The Final Factor*	68
Explore	**INVESTIGATION:** Green Light—Red Light	69
Explain	**READING:** The Three Phases of Stopping	71
Explain	**CONNECTIONS:** How You Can Affect the Three Phases of Stopping	74
Elaborate	**INVESTIGATION:** Your Personal Reaction Time	74
Elaborate	**INVESTIGATION:** Determining Reaction Distances and Perception Distances	79
Elaborate	**CONNECTIONS:** Total Stopping Distances	83
Elaborate	**INVESTIGATION:** Setting Speed Limits	85
Explain	**READING:** The Normal Curve and Setting Standards	87
Elaborate	**INVESTIGATION:** Don't Drink and Drive	89
Elaborate	**CONNECTIONS:** Drinking and Driving—Your Decision	93

CHAPTER 5

Evaluating Your Understanding of Limits and Diversity — 95

Evaluate	**READING:** What Can You Do with Unit 1?	96
Evaluate	**INVESTIGATION:** How Much Noise Is Too Much Noise?	97

UNIT 2

How Does Technology Account For My Limits? — 103

COOPERATIVE LEARNING OVERVIEW — 104

CHAPTER 6

Consumer Concerns — 107

Engage/Explore	**INVESTIGATION:** Tall, Dark, Handsome, Strong, and Absorbent	108
Explore	**READING:** Paper Towel Consumers	111
Explore	**CONNECTIONS:** Comparing Ratings	118
Explain	**READING:** Why Products Fit	118
Elaborate	**CONNECTIONS:** Do You Understand Criteria and Constraints?	124
Elaborate	**INVESTIGATION:** Part of Your Complete Breakfast	124
Evaluate	**CONNECTIONS:** Evaluating Your Understanding of Criteria and Constraints	126

CHAPTER 7

Your Designing Ways — 131

Engage	**INVESTIGATION:** Bon Voyage, Tom Thumb!	132
Explore	**READING:** Is a Boat a Boat?	133
Explore	**INVESTIGATION:** Sails, Propellers, and Gas	136
Explore	**INVESTIGATION:** Anchors Away!	145
Explain	**CONNECTIONS:** Technological Problem Solving	146
Elaborate	**INVESTIGATION:** Toys for Tots	153
Evaluate	**CONNECTIONS:** Human Factors as a Design Constraint	158

CHAPTER 8 — Why Are There So Many Products That Do the Same Thing? — 161

Engage	**INVESTIGATION:** One Problem, Different Decisions	162
Explore	**INVESTIGATION:** Shapely Designs	162
Explain	**READING:** Similarity and Diversity in Designs	167
Elaborate	**INVESTIGATION:** Up, Up, and Away!	169
Evaluate	**CONNECTIONS:** Explaining Design Diversity	173

CHAPTER 9 — Masters of Design — 175

Explore/Explain	**READING:** Let's Talk Technology—Again	176
Elaborate	**CONNECTIONS:** Evaluating Your Environment	177
Elaborate	**INVESTIGATION:** Enabling the Disabled	178
Evaluate	**CONNECTIONS:** What Is Technology	180

UNIT 3 — Why Are Things Different? — 183

COOPERATIVE LEARNING OVERVIEW — 184

CHAPTER 10 — Properties: The Material World — 187

Engage/Explore	**INVESTIGATION:** Diversity of Bounceability	188
Explain	**READING:** How Things Are Different	190
Elaborate	**INVESTIGATION:** Properties for Sale or Rent	196
Evaluate	**CONNECTIONS:** Is the Most Always the Best?	198

CHAPTER 11 — Scientific Explanations Are Ancient History — 201

Engage/Explore	**INVESTIGATION:** Let There Be Colored Light!	202
Explore	**INVESTIGATION:** Strange Phenomena	202

| Explain | **READING:** The Chinese and Greeks Tell What Happened | 206 |
| Elaborate | **CONNECTIONS:** Thinking Like the Ancients | 209 |

CHAPTER 12 — Using Scientific Models to Answer Questions — 211

Explore	**INVESTIGATION:** Mystery Box	212
Explain	**READING:** Another Explanation	214
Elaborate	**INVESTIGATION:** Model Judges	219
Elaborate	**CONNECTIONS:** Particle Movement—Improving the Model	222
Evaluate	**INVESTIGATION:** Judging the New and Improved Version	223

CHAPTER 13 — Using Models to Test and Predict — 229

Engage	**CONNECTIONS:** Model for Sale	230
Explore	**INVESTIGATION:** Gloop	232
Explain	**READING:** More On Models	238
Elaborate	**INVESTIGATION:** Leak-Free Models	242
Evaluate	**INVESTIGATION:** A Penny's Worth of Water	248
Evaluate	**READING:** A Promise Is a Promise	251
Evaluate	**CONNECTIONS:** Properties and Models in Review	253

UNIT 4 — Why are We Diverse? — 255

COOPERATIVE LEARNING OVERVIEW — 256

CHAPTER 14 — You: A Model for Diversity — 259

Engage	**INVESTIGATION:** Taster's Choice	260
Explore	**INVESTIGATION:** Wheel of Traits!	262
Explore	**CONNECTIONS:** Traits and Trees	268
Explain/Elaborate	**READING:** A Model That Explains Diversity	271
Elaborate	**INVESTIGATION:** Too Tall or Too Short for Your Genes?	283
Evaluate	**CONNECTIONS:** Create a Tree	285

CHAPTER 15 Genes and Society — 287

Engage/Explore	**INVESTIGATION:** Designer Reggers	288
Explain	**READING:** Genetic Engineering	291
Elaborate	**CONNECTIONS:** Society and Genetic Engineering	296
Evaluate	**INVESTIGATION:** Science Fiction?	297

How To #1	301
How To #2	309
How To #3	317
How To #4	319

Glossary	325
Acknowledgments	333
Index	335

Preface

Welcome to *Middle School Science & Technology!* We developed this science program specifically for middle school students, such as yourself. In our process of development, we considered middle school students first, then middle school teachers, and then other aspects of middle schools. To develop a program that reflects what middle school students want and need, we had to communicate with many middle school students. We ended up talking with or writing to more than 20,000 students across the United States and in South America. If you look at the credits on p. 333 in the back of the book, you can see where all those students went to school.

We took the following steps to find out what middle school students liked or did not like about science and technology and what we therefore should include in this program:

- We interviewed students about their preferences for science topics. We wanted to know which topics they liked and which they did not. We found out that many topics appealed to middle school students, so we included a great variety of topics in this curriculum.

- We surveyed students to find out which teaching methods they liked and which they did not, because we found out that the preference students had for particular topics depended on how their teacher presented the topic. We found out that very few students liked lectures but that everyone liked *doing* activities. As a result this program depends on your doing and thinking about many activities, and your teacher will be lecturing only a very few times.

- Then we wrote the first draft of the book and let teachers and students try it out for a school year. This trial period is called a field test. As they tried the program, we visited their schools to find out which activities and ideas were working and which were not. Also the field-test teachers and students wrote to us to keep us up-to-date on their progress.

- Using that feedback from middle school students and teachers, we wrote our second draft of the book. More students and teachers then tried it out for another year. This was the second field test. Once again the field-test participants let us know which activities and ideas were going well and which were not.

■ Finally, we compiled all of these comments and reactions to create the book you are about to begin using. This book has combined the input of thousands of students and teachers who took the time to tell us how to improve this book so that you could use and enjoy it.

We hope *Investigating Diversity and Limits* is a program that makes science and technology enjoyable for you to learn. We also hope the activities and ideas in this program make you want to learn more about science and technology. Before long we will be working on the second edition of this program. The second edition will be a revised version of this book. The revisions we make will be based on the feedback of the students and teachers who use this book. If you have comments that you think will improve the second edition for other middle school students, please send your comments to the following address:

BSCS
Attention: MSP
830 N. Tejon Street, Suite 405
Colorado Springs, CO 80903

Sincerely,

Rodger W. Bybee
Principal Investigator

Janet Carlson Powell
Project Director

Program Overview

Middle School Science & Technology is probably unlike other science programs you have used in school. This overview briefly describes some of the key features of the program. You will learn more about these special features of the program when you complete the introductory unit.

Themes

The organization of information in this program might be different from other science programs that you have seen or used. This is because we have organized each level around a unifying theme. One way of describing a unifying theme is as a common idea that keeps coming up to tie other ideas together. Writers use a unifying theme, called a story line, when they write novels. Often music will have a unifying theme that you hear over and over in the piece of music; Beethoven's Fifth Symphony is a famous example of this.

We used a different unifying theme at each level, but each theme has the same purpose: to provide a thread that ties together many scientific and technological ideas. To help keep the unifying theme from becoming repetitive or boring, we use a different curriculum emphasis and focus question for each unit. The curriculum emphasis helped us decide what part of science to emphasize in each unit. The focus question provided a way to connect the unifying theme and the curriculum emphasis to the unit topics in a meaningful way for middle school students. The scope and sequence chart on the next page shows how we organized all of these parts to form an entire program that extends over 3 years. This may be your first year using *Middle School Science & Technology*, or you may have completed 1 or 2 years already. In either case you can use the framework to orient yourself and decide where you have been and where you are going.

The Five "Es"

As you begin using your textbook, you will notice that each page has one or more of the following words at the bottom: *Engage, Explore, Explain, Elaborate, Evaluate.* These words make up five phases that you will use as you learn an idea. These phases describe what you are doing as a learner and what your teacher is doing as a teacher. You will go through each phase approximately once each chapter. For example, when you are doing an **engage**

Scope and Sequence

Level A: Patterns of Change

Unit	1	2	3	4
Curriculum Emphasis	Personal dimensions of science and technology	The nature of scientific explanations	Technological problem solving	Science and technology in society
Focus Question	How does my world change?	How do we explain patterns of change on the earth?	How do we adjust to patterns of change?	How can we change patterns?

Level B: Diversity and Limits

Unit	1	2	3	4
Curriculum Emphasis	Personal dimensions of science and technology	Technological problem solving	The nature of scientific explanations	Science and technology in society
Focus Question	What is normal?	How does technology account for my limits?	Why are things different?	Why are we different?

Level C: Systems and Change

Unit	1	2	3	4
Sub-theme	Systems in Balance	Change through time	Energy in Systems	Populations
Curriculum Emphasis	Personal dimensions of science and technology	The nature of scientific explanations	Technological problem solving	Science and technology in society
Focus Question	How much can things change and still stay the same?	How do things change through time?	How can we improve our use of energy?	What are the limits to growth?

activity, you will be thinking about a new idea. Then you will **explore** that idea in one or several activities. Next you will either **explain** your understanding of the idea, listen to the teacher **explain** more about an idea, or a combination of the two. Then you will **elaborate** your understanding of the idea, usually by doing another activity. Finally you and your teacher will **evaluate** your understanding of the idea. If your evaluation shows that you are successful in understanding the idea, it is time to be engaged in a new idea and to go through these phases for that idea.

Cooperative Learning

We have incorporated cooperative learning strategies into about two-thirds of the program for a variety of reasons. One reason is that cooperative learning gives you a chance to learn and practice how to work successfully with others. The skills that you will gain will become more important to you as you begin to work in an employment setting. Also researchers have shown that cooperative learning can increase the success of students in science class. Cooperative learning also cuts down on the amount of materials needed for each investigation. Your school then can afford to buy other materials so that you can do more investigations. In addition including cooperative learning strategies helps you work like professional scientists and engineers, who do most of their work in cooperative settings in laboratories.

The Characters

We use four cartoon characters in this book (see page xiv). Al, Marie, Isaac, and Rosalind are in the book for four main reasons:

1. To provide a concrete method for demonstrating the value of different learning styles,
2. To make the book friendlier,
3. To teach some of the history of science, and
4. To provide a type of positive role model.

The characters are introduced and described thoroughly in the introductory unit. That information identifies the strengths of each character's learning style, the historical reference for the character's name, and a bit about the character's personality. In the text the characters provide examples of why science is something everyone can do, because they are a diverse group of learners and each contributes something positive to the group.

Questions

There are two primary places you will find questions in this book: in readings as stop and discuss points, and at the end of investigations as Wrap Ups. Some of the students who field tested the curriculum commented that the questions were hard to answer because they had to think about their answers instead of copying

them out of the book. This reaction satisfied one of our goals for this program: to increase your ability to think critically. You will notice that many of the questions in the book have more than one answer that can be considered to be correct. This is because we tried to write questions that you can answer in a variety of ways as long as you provide support for your answer. If this is a new way of answering questions for you, you might feel frustrated at first, but eventually you might find that you enjoy learning this way. Using this questioning method places you more in charge of your own education.

Assessment

Because the topics, themes, and questions in this book are different from most other science programs, it only makes sense that we would include different assessment methods for measuring your success and progress. In many programs you are assessed only by your performance on quizzes and tests. In this program we have recommended that teachers use a variety of assessment strategies that include daily notebooks, checklists, performance tests, short-answer tests, and portfolios to measure how much you have learned and improved and to identify areas that you might want to focus on for future improvement. So don't be surprised if you have "tests" that don't remind you of the tests you are used to taking.

Safety

As in any science program, safety is a concern for everyone who uses *Middle School Science & Technology*. We have made every effort to alert you, the learner, to potentially dangerous situations or materials. We have marked these places with the following symbol:

> ▲ **CAUTION:**

In addition your teacher should tell you about the safe behaviors that you should use in a science classroom. It is your responsibility to follow all safety warnings, rules, and procedures to avoid possible injury to yourself or others.

UNIT 1

What Is Normal?

"It's normal to cry during a sad movie." "You're a normal height for your age." "It's normal for young children to be afraid of the dark." You probably have heard statements like these before, and perhaps you've wondered what people mean by "normal." You even might have wondered whether or not you are normal. If you weren't afraid of the dark when you were a young child, does that mean you aren't normal? If you are five inches shorter or taller than most people your age, does that mean you aren't normal? Is it normal to be the same, or is it normal to be different?

In this unit, you and your classmates will explore your abilities to do various things. In the process, you will be exploring the question What is normal?, and you will use your answer to set standards.

COOPERATIVE LEARNING OVERVIEW

In the cartoon scene on the facing page, the characters are assembling themselves into a cooperative team. It appears that they know something about cooperative learning. Perhaps this is because they worked cooperatively the previous year or perhaps because they have just finished working on the **Cooperative Learning Unit**. In either case, what they are doing merits some study.

First of all, the members of the team have fashioned name tags and role tags for themselves. Your teacher might ask you to do the same, or you might decide to do it as a team. Name tags help your teammates and other members of the class learn your name. Role tags help identify what you should be doing. For example, if you are the **Manager** and you are walking around collecting materials, your teacher and other students can see by your tag that you are behaving appropriately.

The roles you will use in this book are **Communicator, Manager, Tracker, and Team Member**. As Ros points out, you are supposed to assume the role of a Team Member in all cooperative activities. You might also have to assume another role during certain investigations. Refer to the list of role descriptions frequently to make sure you understand the duties of the roles you assume.

Finally, Al mentions that it is time to make a T-chart. During this unit, you will work on one skill during every cooperative activity. The skill is, as Al mentioned, to show caring and respect for others and their ideas. Before you begin the first chapter, you need to discuss this skill and create a T-chart according to your teacher's instructions. You will need to refer to this T-chart to help you practice the skill. Even if the activity does not specifically mention this skill, you should try to practice it.

Each cooperative activity will mention another skill that you should practice in addition to the unit skill.

CHAPTER 1

Identifying Limits and Diversity

Comic book super heroes are not like you and me. Superman has X-ray vision and can fly faster than a speeding bullet. His super powers enable him to perform feats that are beyond the limits of normal humans. But what are your limits? And what is normal?

In this book, we will use the term **limit** to mean boundary, or something that you can't go beyond. For example, you are familiar with city limits, the points at which a city begins and ends. You know about speed limits—driving faster than the speed limit is breaking the law. You might have used a coupon for a free hamburger that said, "Limit one time only." A limit also can be the maximum amount of something; for example, the most people you can fit in a phone booth. Speaking of phones, you might have a limit on how long or how late you can talk on the phone. How many other limits can you think of?

You've probably noticed that you and your classmates have different abilities to do certain things. Some of your classmates can run fast, and others can't. Some of you need eyeglasses to see clearly, and others don't. We will use the term **diversity** to describe this variety in individual abilities. You and your classmates have a diversity of limits to what you can do.

In this chapter, you will explore your limits by performing three different tasks. Then you and your classmates will graph your performance data and look for a pattern.

INVESTIGATION:
Star Tracers

How well do your brain and hands communicate with each other when your eyes aren't getting the messages they usually get? How quickly can you adjust to such a situation? In this investigation, you will attempt a challenging task. You and your teammates will measure how well you each accomplish this task, and you will share your results with the class.

Working Environment

Work cooperatively with a partner. You will need the roles of Communicator and Manager. Each of you will be a Team Member. Check the description of roles to be sure of your duties. Move your desks together so they are facing each other or sit at a table across from each other. Work on the unit skill and the skill Use your teammate's name.

Materials

For each team of two students:
- 1 mirror, 5-by-7 in.
- 1 cardboard box to make a visual shield
- 1 pair of scissors
- stopwatch or clock with a second hand
- 10 copies of the Star Pattern Test Sheet

Procedure: Part A—The Social Skill

1. Prepare your notebook for this investigation.

   Notebook entry: Record the title of the investigation and the date.

2. Write your teammate's first and last names in your notebook.
3. Discuss why it is important to use your teammate's name.
4. Record two of the best reasons you discussed.

Procedure: Part B—The Activity

1. Copy the column headings from Figure 1.1 into your notebook. Copy them across the top of the page so that you can record information from this activity.

   This is called a data table. When scientists conduct investigations, they record their information in this logical, organized way. You will construct many data tables in the future, so this is an important skill

Figure 1.1

Make a data table just like this one on a page in your notebook. You will use the table to record your team's data from Star Tracers.

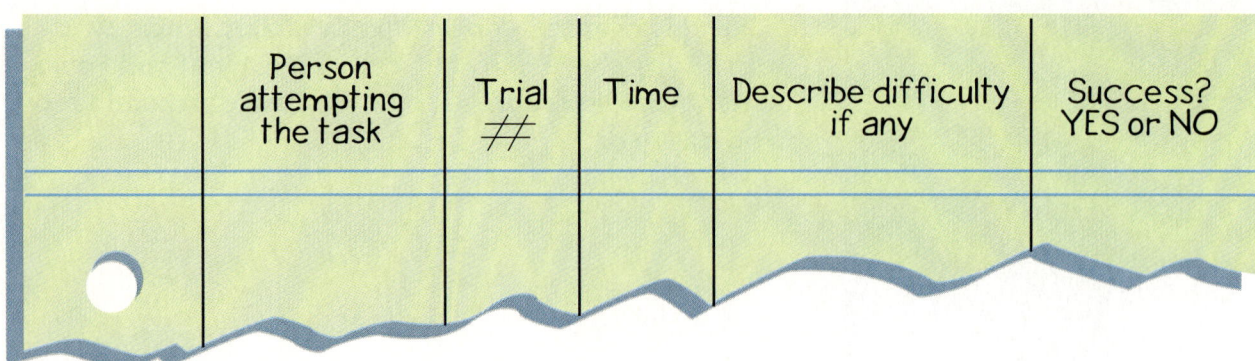

Person attempting the task	Trial #	Time	Describe difficulty if any	Success? YES or NO

6 ■ What Is Normal?

Engage ■ Explore

Figure 1.2

This is the correct way to use your Star Tracers setup.

to learn. *If you want to know how to construct data tables, refer to How to Construct a Data Table, How To #1.*

2. Obtain all the materials you will need for this investigation.

   This is the Manager's role.

3. Cut out two opposite sides of your cardboard box and place the box on the table or desk in front of you like a tunnel.

   This will serve as a visual shield to keep you from seeing what your hands are doing.

4. Sit behind the shield and write your name on a copy of the Star Pattern Test Sheet then slide it under the tunnel.

   The Manager will do this first.

5. Place the mirror in front of the tunnel and position it so the person doing the tracing can see his or her own hands and the star in the mirror.

   The Communicator should position the mirror for the Manager, who will be the first tracer. You will take turns doing this activity. The person doing the tracing should not be able to observe his or her hands directly (see Figure 1.2).

6. When your teammate is ready to trace, say "Begin," and start timing.

   The Communicator is the first timer and starts timing by noting the starting time on a clock or pressing the start button on the stopwatch.

7. When you hear the word "Begin," use a pen or pencil to start tracing the star pattern between the two borders of the star.

   Watch your hands in the mirror.

8. When you complete your first attempt at tracing the star, say "Stop."

   The Communicator should note the stopping time on the clock or stopwatch.

9. Calculate the total time it took your teammate to trace the star.

   The Communicator will do this.

   STOP: Are you using each other's names?

10. Fill in the columns of your data table with the appropriate information from this tracing.

    Together with your teammate, you will need to decide what you will count as an error and how you will determine whether or not each tracing was a success.

11. Attempt to trace the star four more times, using a new test sheet each time.

    The Manager traces the pattern four more times. After each attempt, the Communicator records the necessary information under the appropriate headings in the data table.

12. After five tries, trade places with your teammate.

    The Communicator now tries to trace the star five times, while the Manager records the appropriate information on the data table.

13. Analyze your data.

    This task should be simpler because you have organized your data into a data table. To analyze your data, do the following: (a) count the number of times each person was successful in tracing the star, (b) count the number of times each person was unsuccessful in tracing the star, and (c) calculate the total successful attempts and total unsuccessful attempts for your team.

14. Record each of these totals.

    Notebook entry: Record this in your notebook following your data table.

Wrap Up

Write the words "Wrap Up" in your notebook. Discuss the following questions with your Team Members and write your answers in your notebook. Be certain you can explain your answers during a class discussion.

1. Did anyone on your team trace the star successfully the first time he or she tried the task?
2. Describe your experience of trying to trace the star by looking in a mirror.
3. What limits did you experience as you tried to trace the star?
4. As a team, decide how successful you were at using each other's names: excellent, good, fair, or poor.

INVESTIGATION:
Threading the Needle

In Star Tracers, your ability to draw a star was limited because you couldn't see your hands. Some of your classmates may have been quite successful, while others were probably less successful. In other words, there was a diversity of limits within your class. Will the same people who were very successful in Star Tracers be successful in another investigation that places a different limit on their vision? This investigation gives you a chance to find out.

Materials

For each team of two students:
- 1 eye bolt
- 1 bolt
- 2 rulers
- 2 sheets of graph paper

Procedure

1. Construct a data table to use with this activity.

   Follow the steps in How To #1, How to Construct a Data Table. Each person will need a data table in her or his notebook.

2. Collect the materials you will need for this investigation.

   This is the Manager's role.

3. Stand up. Hold the eye bolt in one hand at arm's length in front of you with your elbow slightly bent. Hold the bolt in your other hand with your arm down at your side.

Working Environment

Work cooperatively in your team of two. Use the roles of Communicator and Manager as well as Team Member. Continue to practice the unit skill, and the skill Use your teammate's name. Push your desks together, side by side, or sit beside each other at a table. Clear enough space by your desks or table so you can stretch out your arms.

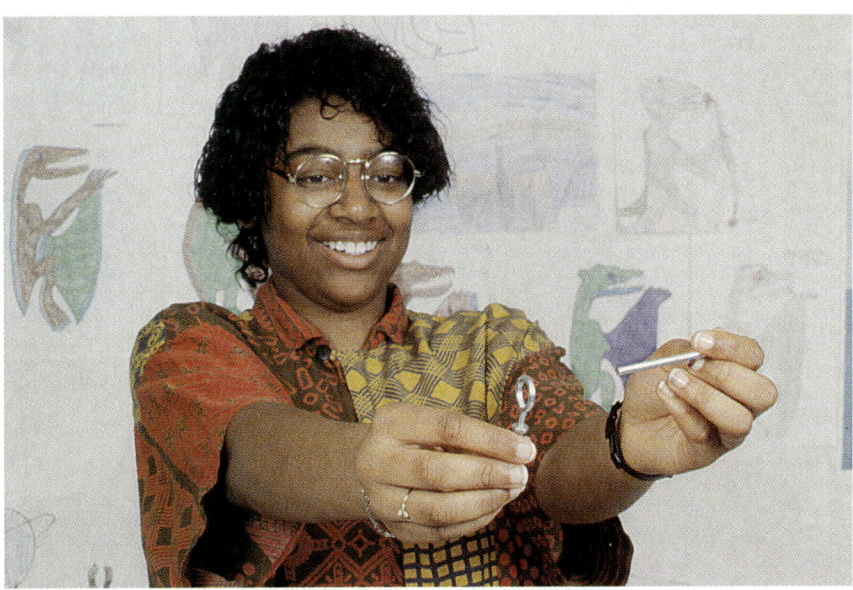

Figure 1.3

This is one way to thread the eye bolt.

The Communicator will do the task first, while the Manager records the results of each attempt in your data table.

4. Close your right eye. Try to get the bolt through the hole of the eye bolt in one motion.

   You have five tries.

5. Close your left eye. Try to get the bolt through the hole of the eye bolt in one motion.

   You have five tries.

6. Keep both eyes open. Try to get the bolt through the hole of the eye bolt in one motion.

   You have ten tries.

7. Record which attempts were successful and which were not successful for each set of attempts described in steps 4 through 6.

   Notebook entry: Record this in your data table. Help your partner by recording data for him or her.

8. Switch places and repeat steps 3 through 7.

   Don't forget to record the data.

9. Exchange and record the information in your data tables.

   Each person should have all your team's attempts recorded in his or her own data table.

10. Return your materials to the appropriate location.

11. Analyze your data.

    Calculate these totals for both people: (a) the number of successes with the right eye closed, (b) the number of successes with the left eye closed, (c) the number of successes with both eyes open, and (d) the total number of successes.

    Notebook entry: Record this information in your notebook following your data tables.

12. Enter each Team Member's individual success on the class data table.

    Your teacher will provide this on the board or on an overhead transparency.

Wrap Up

Write the words "Wrap Up" in your notebook. Discuss the following questions with your team and record your answers. Be sure to write each answer in complete sentences and be able to explain your answers.

1. Did you need to change your data table during this activity?
2. Did the data table make it easy or difficult to analyze the data?
3. Compare how many times you were successful with either eye closed and with both eyes open.

If I do it with both eyes closed, that's a new limit for myself!

4. Describe the limits that influenced your ability to perform this task successfully.
5. How much diversity was there between your ability and your partner's ability to put the bolt through the eye bolt?
6. Describe how closely your team's number of successful attempts matches the success rate of other teams in your class.
7. Review the rating your team decided on for how well you practiced the skill of using your teammate's name. (Recall that this is the rating you gave in the Wrap Up to Star Tracers.) How would you change that rating for this investigation?

 **CONNECTIONS:**
How Do You Spell Success?

Discuss your answers to the following with the rest of the class.

1. Describe how your team determined which team had the most success during the Star Tracers investigation.
2. How many different definitions of success were there in Star Tracers?
3. How did each team define success in Threading the Needle?
4. Describe whether you think it is fair to compare all the teams and say that the team with the most successful attempts in Star Tracers or Threading the Needle is the best team in the class.
5. Develop a class definition of a successful attempt at threading the bolt through the eye and record that definition in your notebook. Remember that you will need to account for both technique and your definition of what is a success.

 **INVESTIGATION:**
If at First You Don't Succeed

In this investigation you will use your class definition of success to repeat the Threading the Needle investigation. You will use the same materials and basic procedure, except you must adhere to class decisions. The object is to do everything the same way as everyone else.

Materials

For each team of two students:
- 1 eye bolt
- 1 bolt
- 2 rulers
- 2 pieces of graph paper

Working Environment

Work cooperatively in your team of two. In addition to Team Members, you will need a Communicator and a Manager. Set up your working environment as you did in Threading the Needle, but this time practice the skill Move into your groups quickly and quietly.

Procedure: Part A—The Social Skill

1. Construct a T-chart with two columns.
2. Label one column "Sounds Like" and the other column "Looks Like."
3. Discuss the social skill for this activity and then, as a team, fill in the columns based on your discussion.

Procedure: Part B—The Activity

1. Create a data table.

   You may use the format for Threading the Needle or try a new data table format if your last one didn't work well for you.

2. Refer back to the procedure for Threading the Needle and repeat steps 3 through 11.

   This time you must use the class definition of success from the previous connections section to complete the task and to measure success in your team. Only repeat steps 3 through 11. Do not do the Wrap Up for Threading the Needle again.

3. Enter your new data on the class data table.
4. Construct a graph of the new class data.

   Refer to How To #2, How to Construct a Graph, if necessary.

5. Compare this class graph you made to the class graph your teacher made for Threading the Needle.
6. Help your teacher construct a graph of the class data for this investigation.

   Use your graph and offer suggestions. Correct your graph if necessary.

Wrap Up

Discuss the following as a team. Record your answers in your notebook. Be ready to discuss the questions with the rest of the class.

1. Did you change the format of your data table? If so, describe what changes you made and why.
2. This time did your number of successes in threading the eyebolt differ from your partner's? If so, how do you explain the difference?
3. List any new limits you think were placed on you by how the class defined success.
4. Describe the differences between the first class graph and the second class graph.
5. Have the heights of the bars and the differences between the curve over the bars changed?
6. Which graph shows the most diversity among your classmates?

7. Decide which graph gives you the most reliable information about class success in threading a bolt through an eye bolt and explain why you think the graph you selected is more reliable.
8. On a scale of 1 to 10, 10 being the best, rate your team on the use of the social skill for this activity.
9. List specific strategies that your team could try in order to improve moving into your group more quickly and quietly.

READING:
Doing It All the Same

When you first measured success at different tasks, chances are you agreed with Al's approach. Now that you have tried to compare the results of people who do things differently, however, you might

agree with Isaac. Of course, you probably didn't get as carried away with defining a successful measurement as Isaac did, but you eventually did agree with your classmates on how you would measure success.

The different class graphs show that when you decide on a similar way of doing things, it makes a difference in your results. It also, then, shows a difference in the pattern of the graphs. In fact, the graph that provided the most reliable information was the one you constructed after you agreed on a class definition of how you would measure success. This was because you eliminated the differences in the techniques that you and your classmates used to thread the eye bolt, and you had agreed on what you could consider a success. By eliminating the different techniques, you tested each student's ability to accomplish the task within given limits. The first time you completed Threading the Needle, the members of team A might have held the eye bolt at their side, while the members of team B might have held the eye bolt in front of their bodies. The difference in the success rate between team A and team B might result from their different abilities, but it may be due to the fact that they held the eye bolt differently. Without keeping the position of the bolt consistent, you have no way of knowing for sure.

To be certain that your results reflect only the differences you wish to test, you must consider all parts of your experiment that can change or "vary." This includes the position of the bolt and the distance of the bolt from your body. Anything that can affect the results of your experiment is called a **variable.**

You can **control variables** in an experiment by keeping all of the variables constant except for the one that you want to test. When you do this, you can then compare your results with others that do the same experiment. In order to do this, you must decide how to keep all variables the same for everyone doing the experiment throughout the experiment. For example, in Star Tracers, the class might decide that each team should remain silent during an attempt and everyone should arrange their desks so the amount of light is constant from team to team. Then, the only variable left would be each student's ability to trace the star. That is exactly the variable you wanted to test.

Stop and Discuss

1. What variables were affecting your measurement of class success the first time you did the eye bolt activity?
2. Which variables did your class decide were important to keep constant?
3. How did you control variables the second time you did the eye bolt activity?

One way to control variables is to agree on one class technique. But, we still have more work to do. Think again of the star tracing experiment. One team might have allowed its members to cross the border of the pattern in a successful attempt. Another team might have required its members to complete the tracing without crossing the border at all. The second team would have fewer successes. Without a standard definition of how to measure success, you might never make a fair comparison between teams. With a standard definition, each team will measure success in exactly the same way.

A standard definition of how you measure something is called an **operational definition.** You can develop operational definitions of techniques for most scientific investigations. Once you do so, someone else can repeat the experiment and make fair comparisons to your results.

Stop and Discuss

4. Read the following operational definitions and decide which definition is the best one to use to measure success in Star Tracers.

   a. You are successful in tracing the star pattern when you draw the star pattern well.

   b. You are successful in tracing the star pattern when you draw the star pattern without crossing either border.

   c. You are successful in tracing the star pattern when you draw the star pattern in one minute without touching either border.

5. Explain why you chose the operational definition you did.

6. How would you explain to a friend what an operational definition is?

Identifying and controlling variables and deciding on operational definitions are important skills in conducting investigations. Given time and practice, your ability to use these skills should become second nature to you.

INVESTIGATION:
Seeing the World around You

During the previous investigations, you explored diversity among students in your class as they accomplished small tasks. You discovered that people have different limits in accomplishing certain tasks. You tried your hand at recording and analyzing data in data tables. You made a graph of your data that showed your results. You learned about controlling variables and using operational definitions.

This investigation will bring all of this knowledge and these skills together. You will construct data tables, graph data, and explore the limits of peripheral vision. Peripheral vision is the ability to see around you without moving your head. Because all of you will test your peripheral vision and will want to compare your results, be certain to make the investigation fair. That means use a common operational definition and control variables.

Working Environment

Work cooperatively in your teams of two. One person will be the Manager. The other will be the Communicator. Push your desks together. Try to be the group who moves most quickly and quietly.

Materials

For each team of two students:
- any materials you need from those your teacher provides
- graph paper

Procedure

1. Construct a data table for this investigation.

   Remember that means reading through the entire procedure first.

2. Determine what procedure you will use for measuring how far you can see around your head without moving your head.

   You can measure this any way that you want, but you will have to compare the results of all the teams in your class. This measurement is called peripheral vision: Read the Background Information following this procedure to learn more about what you will measure. As you decide on your procedure, consider the following:

   - *If you have to compare results, you will have to make the test fair. What can you do to ensure a fair test?*
   - *If you have to compare results, you will have to think of a way in which you will compare the same kind of data. How will you make sure you are comparing similar results?*

   STOP: Are you showing caring and respect for the ideas of your Team Members?

3. Visit the materials station, carefully look at what is available, and make a list of these items.

   This is the Manager's role.

4. Decide what materials you will need.

   Both Team Members should review the list and decide together.

5. Experiment with the materials to figure out a way to measure peripheral vision and to see what affects peripheral vision.
6. Conduct your investigation.

    Notebook entry: Record your data in your data table.

7. Record your data on the class data table.
8. Construct a class graph to show the number of people that have each specific peripheral vision measurement.

    Use your class data table.

Background Information

If you look straight ahead and see something moving to one side of your head, you are seeing that object with your peripheral vision. When you see just a shadow or a movement, but you cannot identify the object, it is near the outer limit of your peripheral vision. The point at which you can first identify the object (a red pencil, a blue car, your mother) marks the beginning, or inner limit, of your peripheral vision. If the object kept moving around your head to the front and then to the other side, it again would reach a point where you could no longer identify it and then a point where it would disappear. These points mark the inner and outer limits of your peripheral vision **range**. A **range** defines the inner and outer limits of a characteristic such as peripheral vision. Ranges exist in individuals as well as within populations.

Wrap Up

Write the words "Wrap Up" in your notebook. Discuss the following with your partner and write the answers in your notebook. Be sure to use complete sentences.

1. Describe how well your data table worked for this investigation.
2. Do you and your partner show diversity in your peripheral vision limits?
3. Describe the operational definition your class used to measure peripheral vision.
4. Describe which variables may have caused a difference in results among teams.
5. Look at the class graph and decide where these points would be: (a) the smallest value recorded for peripheral vision, (b) the largest value recorded for peripheral vision, and (c) the top of the curve.

6. Based on the class graph, answer these questions:
   a. What is the range of peripheral vision measurements on the graph?
   b. Do most students have a peripheral vision measurement greater than or less than the measurement at the top of the curve?
   c. Describe the diversity of peripheral vision measurements in your class.

7. How much have you improved your use of the unit skill: greatly, slightly, or not at all? Describe one thing you will do in Chapter 2 to keep improving your use of this skill and one strategy to improve moving into your groups quickly and quietly.

READING:
The Human Eye and Peripheral Vision

People often say, "I saw it out of the corner of my eye!" The images you see from the corners of your eyes often don't tell you much, but they do encourage you to turn your head and take a better look. This "corner vision" is the peripheral vision you just investigated. In order to understand how it works, take a closer look at the eye.

If you could remove your eyeball from its socket, it would feel like a ball of jelly enclosed in a tough case. This case is opaque and white except at the front where the clear cornea is located. If you could cut the eyeball in half with a vertical slice through the middle, you would see something like the picture in Figure 1.4.

Behind the cornea is the iris, or the colored part of the eye. The iris is a muscle that opens and closes to let more or less light in. Directly behind the iris is the lens, which focuses images. The rest of the eye behind the lens is filled with a jellylike material. The layer on the inside of the eyeball is called the retina. The retina is composed of an amazing network of cells shaped either like cones or rods. Among other functions, the cone-shaped cells help you see colors and the rod-shaped cells help you see in dim light.

Figure 1.4

The eye is composed of three layers that form a fluid-filled sphere. The front part of the sclera is called the cornea and it is transparent. The colored part of the eye, the iris, is a muscle that opens or closes to control the amount of light coming into the eye. The pupil is a hole in the center of the iris. The transparent lens helps the eye focus. The innermost layer, called the retina, contains the light-sensitive cells, the rods and cones.

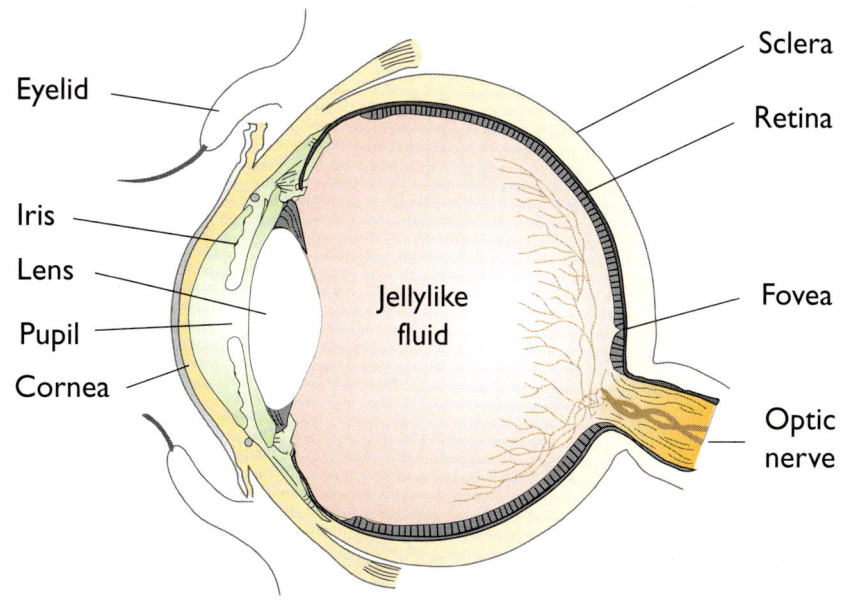

Light from an object passes through the lens and reaches the retina, where it falls onto the cones and rods. Most rods and cones are located in the middle of the retina and only a few are located around the edges. In other areas of the retina, the cones and rods aren't packed together so tightly. If the light hits the part of the retina that contains the most rods and cones, called the fovea, you are able to identify the object.

Light entering your eye from the side does not fall on the fovea, but on the part of the retina where there are fewer rods and cones. You are less able to identify images that result from light coming into your eye from your sides. You are better able to identify images that result from light coming into your eyes directly in front of you. This explains why peripheral vision is limited. The placement and numbers of cones in your retina limits how well you see colors in your peripheral vision. The placement and numbers of rods and cones differ among people. This accounts for the fact that we have a diversity of limits in our peripheral vision.

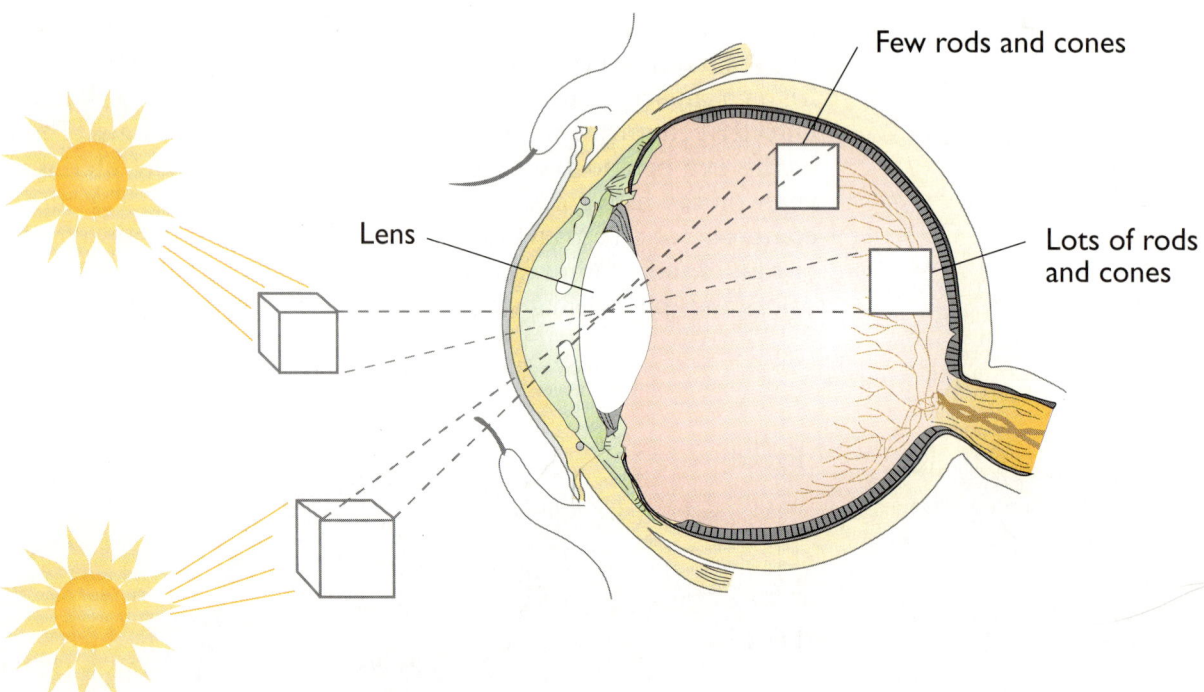

Figure 1.5

When you see an object directly in front of you, the light from this location enters the eye and falls on the fovea, which consists of closely packed rod and cone cells. This results in a sharp, crisp image. When you see an object off to your side, the light falls on the part of your retina that has few rods and cones. The images that result are not as clear as the images of objects directly in front of you.

SIDELIGHT

Limits and Diversity in Animal Senses

Some animals have limits on their senses that are much different from humans. Some have poor hearing or eyesight, while others have special adaptations that give them extraordinary abilities. How do these animals adjust to these limits?

Although bats have poor eyesight, they are not blind. They "see" in a different but very effective way. As a bat flies, it constantly makes high-pitched sounds. As these sounds travel away from the bat, they bounce off objects. The sounds then come back to the bat. When the bat hears sounds coming back, it knows that there is an object ahead. This ability to use sound to locate objects is called echo-location. Using echo-location, bats hunt for food and find their way at night. Echo-location is so good that bats can fly higher, faster, and farther at night than most other animals that fly at night.

Other animals with different limits to their senses include the following:

- Rattlesnakes and pythons can see the infrared radiation, or "heat," given off by the bodies of living animals.
- Goldfish can see not only infrared but also ultraviolet radiation. (Human eyes cannot detect either of these types of radiation.)
- Frogs have special types of cells in their retinas that trigger an automatic reaction when anything "fly-sized" moves in front of them. Frogs often respond only to things that are of this size. Much of the visual information that people record may never even reach a frog's brain.
- Flies have the ability to take in many separate images rather than a single image. They see the world as though they are looking through a kaleidoscope. This helps them survive in their high-speed world and track the motions of their predators.

Elaborate

CHAPTER 2

Normal Ranges of Limits and Diversity

In Chapter 1 you and your classmates investigated your abilities to accomplish certain tasks. When you compared your limits, you discovered that your class displayed a wide range of diversity. For example, some people could see farther around their heads than others. If you had a graph of the peripheral vision limits of all the students in your school, do you think it would show as much diversity as you found in your class, or is your class unusual in its diversity? Take a moment to discuss this question. Then you can investigate diversity in something that's probably very familiar: POPCORN!

INVESTIGATION:
The Individuality of Popcorn

In this investigation you will observe imaginary alien creatures from the planet Popcora. You will decide how normal these creatures are.

Working Environment
Work individually as you follow the procedure, then work as a class during the wrap-up section. Your teacher will let you know where to work.

Materials

For the entire class:
- bowl of alien creatures from the planet Popcora
- 1 alien life vessel
- small amount of polyunsaturated lipid
- 30 beakers (250-mL)

Procedure

1. Listen as your teacher reads you the following guided imagery.

    Recently, planet Earth experienced a peaceful invasion of aliens. The aliens landed on the kitchen counter of the White House in a small vessel resembling a salad bowl. Five seconds later, another vessel, more complicated in structure, landed beside the bowl. This vessel resembled a small electric appliance with a base and a lid. Inside the vessel was a message. The message read:

    "We are from the planet Popcora. Our civilization was at war, and just before our planet exploded, we managed to launch ourselves in our life vessel. We were unable, however, to launch our transferring mechanism. We set our destination for your planet, which we have named "Butter." Through our studies, we know that you can help us by providing the proper atmosphere in our life vessel. The proper atmosphere will be achieved with the addition of a small amount of a polyunsaturated lipid substance. Once our life vessel contains the proper atmosphere, transfer us into the life vessel and supply an electric current to provide us with a comfortable temperature. We will then be able to fulfill our prime directive: to shed our protective suits with much jumping and exploding and to be consumed as nourishment by the peoples of many lands and nations."

    Scientists the world over have been studying these creatures called "Popcorn." After numerous inconclusive studies, top government scientists have shipped the Popcorns to your scientific community, which has a reputation for conducting the finest scientific investigations in the world. You have, therefore, been chosen by United States officials to make the final observations of the Popcorns and to help them fulfill their prime directive.

2. Fill your beaker with creatures until the bottom is covered.

    Do this when your teacher has finished reading and then return to your work place with your beaker.

3. Observe your creatures closely and describe their basic features.

   Stir the Popcorns up. Let them fall between your fingers and look closely at them.

   Notebook entry: Be sure you record your observations and answer these questions:

   - *How similar are they to one another? How different are they?*
   - *If each alien had a different name, would you be able to tell one from the other? Why or why not?*
   - *Would you say that these creatures all look normal for Popcorns? Do any not look normal? If so, how many do not look normal?*

4. Provide an atmosphere and some heat to the life vessel and transfer the Popcorns to it.

   The oldest scientist in the room should do this.

5. Carefully observe what happens to the Popcorns as they shed their protective suits.

   All scientists in the room should do this.

   ▲ **CAUTION: The popcorn popper and oil will be very hot. Be careful not to touch either the oil or popper at this point. Unpopped kernels also might be hot.**

Engage ■ *Explore*

Normal Ranges of Limits and Diversity ■ **25**

6. After the Popcorns shed their suits, the senior scientist will distribute some to each group.

7. Record your observations about the aliens and their changes.

   Notebook entry: Be sure you include comments about the size, color, and texture.

Wrap Up

Answer the following questions in your notebook. Be ready to share your answers with the rest of the class. Remember to show caring and respect for your classmates during the discussion.

1. Using your recorded observations, rate the characteristics of the unpopped kernels of popcorn from 1 to 5 according to this scale: (1) all are similar, (2) most are similar, (3) some are similar, (4) a few are similar, (5) none are similar.

a. color	1	2	3	4	5
b. shape	1	2	3	4	5
c. size	1	2	3	4	5
d. texture	1	2	3	4	5
e. smell	1	2	3	4	5

2. Use the same scale and characteristics to rate the popped popcorn and the kernels that didn't pop.

3. Did all of the kernels pop at once?

4. If the kernels popped at different times, are all the kernels normal? Explain your answer.

5. Are the very first and very last kernels of corn to pop normal?

6. If some of the kernels did not pop, are they normal?

7. Explain what normal means to you.

READING: What Is Normal?

In the Star Tracers investigation, some of you were more successful at tracing a star pattern than others. When you tried to thread the eye bolt with a bolt, some of you were more successful at placing a bolt through an eye bolt than others. Also, during the peripheral vision experiment, you discovered that your class had a range of abilities to see to the side. You plotted your data from the peripheral vision investigation and observed the pattern that your data created. On your graph, you indicated the range of peripheral vision for your class by marking the lowest value, the middle point, and the highest value you measured. Your graph is one way to show the diversity of peripheral vision in your class.

Recall the question from the beginning of this chapter: If you had the peripheral vision test results of all the students in your

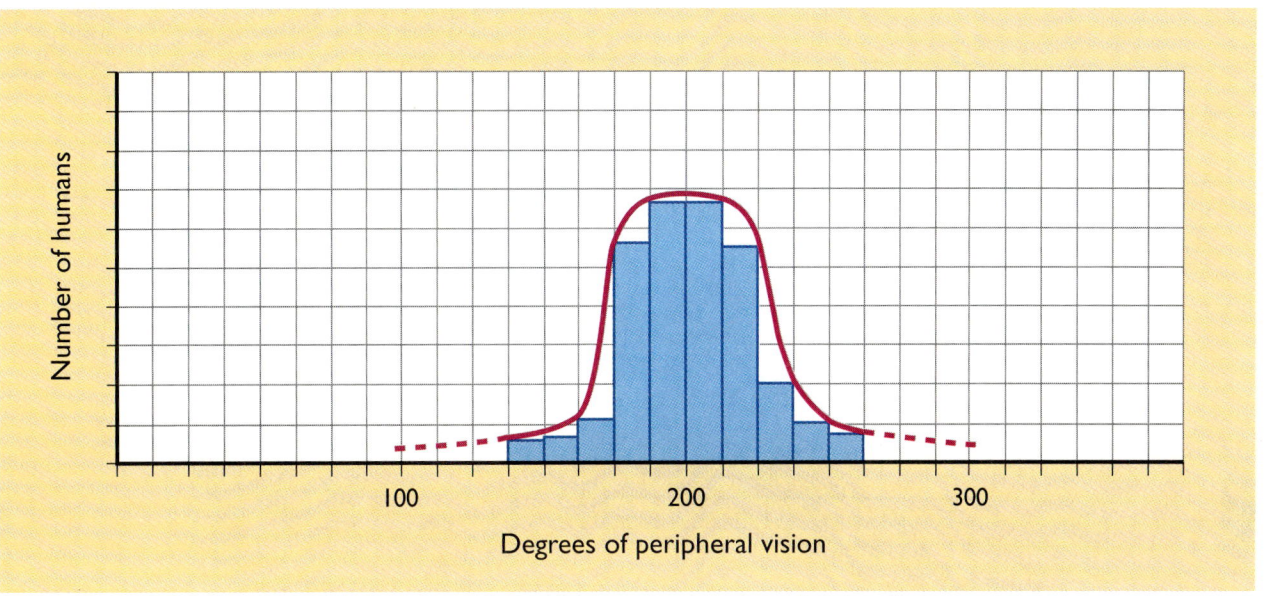

Figure 2.1

This is an example of a graph of people's peripheral vision measurements. The graph of your class data is probably very similar. Most people probably have peripheral vision close to 200 degrees, but there is a lot of diversity. Notice how the data form a normal curve.

school, would you expect to find as much diversity as you found in your class, or are your classmates more diverse than any other class in your school? If you tested everyone in your school, you probably would find as much diversity as you did in your class, if not more. A graph of data from your entire school would be similar to the class graph. In fact, if you tested everyone in the world for peripheral vision, you would still come up with a graph that looked something like the graph in Figure 2.1.

Because the graph of human peripheral vision includes more data, the shape of the curve is probably smoother and more "bell" shaped than your class graph. Otherwise, the graphs are very likely about the same. The graph of human peripheral vision shows diversity, just as your class graph does. The end values on the curve of the graph define the range of human peripheral vision. The bump indicates where the values for most humans' peripheral vision would fall.

Like the popcorn kernels that popped first and last, the person at the far right end of the graph and the person at the far left end of the graph are not *abnormal* just because they aren't where *most* of the people are. Consider the popcorn questions you answered in the previous wrap-up section. Just because some popcorn popped before or after the majority of the popcorn, this doesn't mean that those kernels are not normal. One kernel normally pops first and one normally pops last. Similarly the person with the lowest value and the person with the highest value on the graph of human peripheral vision are normal people. Their measurements comprise the edges of what is known as the **normal curve.**

In the case of human peripheral vision, the normal curve shows us that most people in the world have a peripheral vision range between 180 degrees and 220 degrees, yet some people have

peripheral vision that is above or below that. Where *you* fall on that curve describes your own personal limit. The entire human peripheral vision curve shows the limits in peripheral vision for the entire human population. It also demonstrates that there is a range of diversity in the entire human population. The bump on the

Figure 2.2

In this graph of peripheral vision of other animals, notice that each species shows diversity in its limits. Which animals have peripheral vision most similar to humans? Which animals can see the farthest around their heads?

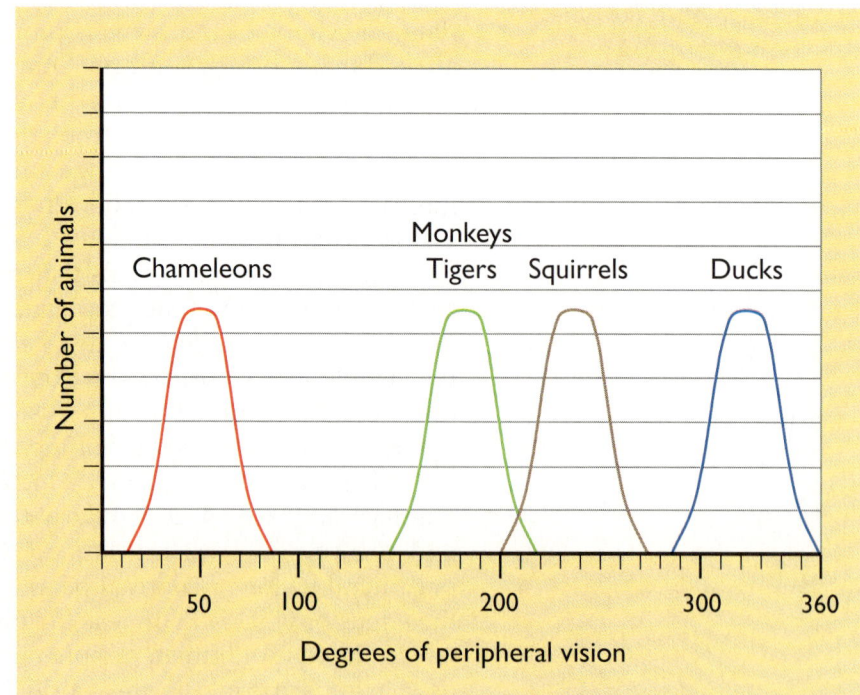

28 ■ What Is Normal?

Explain

graph shows the values for the majority of people's peripheral vision. The highest and lowest measurements mark the beginning and ending of the **range** of possible measurements.

By graphing the data, we emphasize the range of diversity. All the data under the curve represent what is considered normal. If you included some other animals' peripheral vision curves on the same graph, you would see something like the graph in Figure 2.2.

Other animals exhibit a wide diversity of limits. So diversity is found not just in your class but in entire populations of humans and other animals.

What is normal? Diversity is.

INVESTIGATION: A Diversity of Popcorn

In this investigation you will use popcorn again to further your understanding of the normal curve. When you have completed the investigation, see whether you can describe the shape of the normal curve and what it means.

Working Environment

Work individually. You need a working space large enough for a big piece of drawing paper. Your desk top is fine. If you're working at a table, just leave plenty of room between you and other people. You need some standing room beside your desk.

Materials

For each student:
- 1 marker, any color
- 1 large desk-sized piece of drawing paper

For the entire class:
- enough popcorn and oil to pop two batches
- popcorn popper

Procedure

1. Obtain your materials.

⚠ **CAUTION:** The popcorn popper and oil will be very hot. Be careful not to touch them.

2. Squat as low as you can beside your desk.

   Your teacher will pop the first batch of popcorn. Think of yourself as a measuring device, something like a thermometer.

3. Indicate the amount of popcorn popping at one time by adjusting your height according to the sounds of the popping.

   When no popcorn is popping, you should be at the lowest point, which is squatting. As the pops begin, you should slowly rise to your feet and match your height to how many pops you hear at the same time. When the most popcorn is popping, you should be at your highest point, or standing on tiptoe. Lower yourself back to a squat as the number of kernels popping diminishes.

4. When you have completed step 3, sit down and spread your drawing paper across your desk. Hold your marker at the lower left corner of the page and prepare to draw. Close your eyes.

   Keep your "body gauge" experience in mind (what you did in step 3). Your teacher will pop the second batch of popcorn.

5. Keep your eyes closed. As the popcorn pops, draw a line that illustrates what you hear.

   Listen carefully to the sounds of popping. Let the line you draw be the popcorn gauge just as your body was for the last batch. Start drawing your line at the sound of the first pop and finish at the sound of the last pop.

6. Open your eyes and look at the line you drew. Then look at other students' lines.

Wrap Up

Write answers to the following in your notebook.

1. Describe the shape of the line you drew.
2. Explain how other students' drawings are similar to or different from yours.
3. Choose the word or phrase that describes the amount of diversity in your classmates' drawings: a lot, some, none. Why did you choose that word or phrase?

CONNECTIONS:
More on the Meaning of the Normal Curve

With your classmates, help your teacher construct a graph on the chalkboard that represents popcorn popping. Label and number the axes and discuss where to place the high and low values. Your class also should decide where to place the center (highest) point of the curve.

Your teacher will make four sections on the graph, labeled A, B, C, and D divided by three lines labeled 1, 2, 3. Individually, study those lines and sections and agree on answers to the following questions. Write your answers in your notebook and prepare to explain your answers to your classmates.

1. Which line drawn through the curve matches where most of the popcorn was popping simultaneously?
2. In which section(s) of the graph did the majority of the popcorn pop?

3. In which section(s) did the minority pop?

4. Is there a great difference between the amount of popcorn that popped before the high point on the graph and the amount of popcorn that popped after the high point on the graph?

5. Look again at the line you decided represented where the most popcorn was popping simultaneously. Estimate the percentage of the total amount of popped corn this point represents. Estimate the percentage that sections B and C represent together. How do the percentages compare?

READING:
The Value of the Normal Curve

We call the graphs you have been constructing **normal curves.** Normal curves, also known as bell curves, show the normal diversity in various limits for a group of organisms. If you looked around your classroom, you would see a diversity in the height of the students. If you made a list of every person's height, added the heights together, and divided by the total number of people on your list, this would give you the **average** height for students in your class. You could find the **range** for your class by determining the shortest and tallest measurements you recorded. The range would be these two points and everything in between. If you were curious, you could even find the average height and the range of heights for all students in your grade level or school using the same method. Or you could use this same method to determine the average height and height range of all students in your grade level in your city, state, or even the entire United States.

Stop and Discuss

1. You would have a fairly long list of heights just by figuring out the average height and height range for your class. Can you imagine how long your list of heights would be if you tried to figure out the average height and the height range for all the students in your grade level or in the United States?

To avoid adding up an incredibly long list and then dividing, you could determine the average height and height range for students in the country by taking small samples of heights of students your age from different places in the country. Then you could make a data table, record your information, and then plot all the data on a bar graph. The vertical axis would be labeled "Number of students" and the horizontal axis would be labeled "Height in cm." The data would produce a normal curve. On that curve you could draw a vertical line down the middle from the

Figure 2.3

This normal curve shows the length of jeans worn by twelve-year-olds. If you were a clothing manufacturer, in what lengths would you make the most jeans?

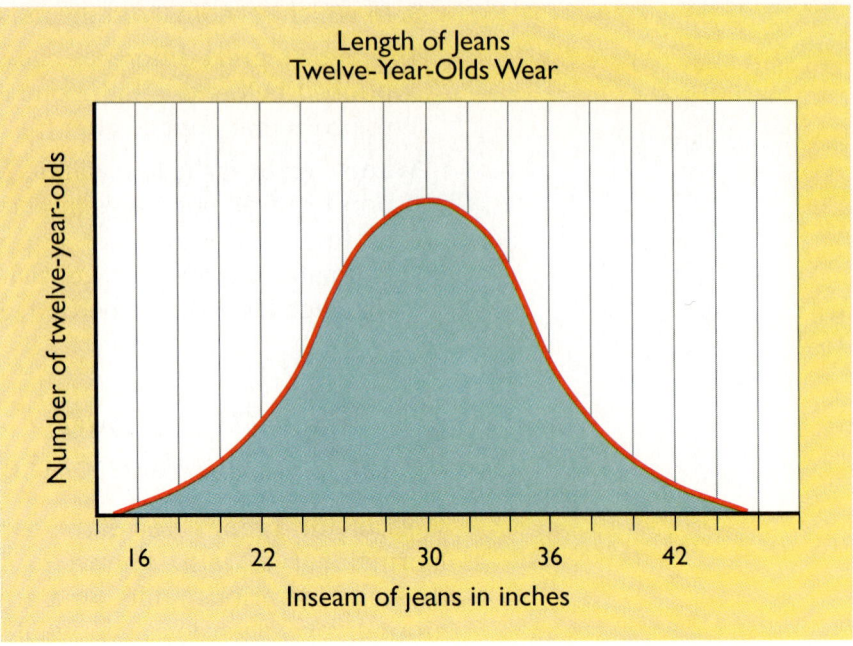

high point of the bump to the horizontal axis. The value on the horizontal axis at that point would tell you the average heights of students in your grade level in the U.S. without having measured every individual student in the country. The end points of the curve would tell you the range. That's the benefit of the normal curve—it provides a lot of information without a lot of work.

Breathe a sigh of relief. You now know how to determine the average height and the range of heights of students in your grade in this country with less work than if you'd used the add and divide method. So what? Who cares, anyway? Actually you might not realize it, but you do. You see, clothing manufacturers can use data from a normal curve to decide on clothing sizes. They can construct a normal curve for the lengths of jean inseams that students your age would wear. They then can use the information from this curve to manufacture the appropriate number of jeans in each size. A curve showing the inseam lengths of all twelve-year-olds in the country might look like the graph in Figure 2.3.

Stop and Discuss

2. What information does the graph contain?
3. List two ways you could make use of the information on the graph.
4. How could clothing manufacturers use this graph to help them make clothes?
5. If you were a clothing manufacturer would you make as many pairs of jeans with a 20-inch inseam as you would with a 30-inch inseam? Why or why not?

Now you have seen how you can apply a normal curve to a lot of different situations: peripheral vision, popcorn, height, and manufacturing clothes. The normal curve is so useful that you can apply it to everyday life as well as to investigations in science class. Data from many other everyday situations often produce a normal curve: shoe sizes, weights, family sizes, number of hours you sleep at night, number of pets at home, number of hours you watch TV per day, and the list could go on! Using the normal curve can be valuable and fun. You will practice using the normal curve in the next investigation.

INVESTIGATION:
The Normal Curve and You

This investigation will give you a chance to find out more about some characteristics of your classmates. You will choose a common characteristic and collect data from your classmates. You might choose to investigate something such as foot length or what times people have breakfast. Then you will plot these data on a graph.

Working Environment

Work cooperatively with your partner. Use the roles of Communicator and Team Member. Push your desks together facing each other, or sit at a table facing each other. Practice the social skill Move into your groups quickly and quietly. As you move into your groups, use the strategy you described in the last wrap-up section of Chapter 1.

Materials

For each team of two students:
- 1 sheet of graph paper
- tape, stapler, or glue

Procedure

1. Decide together what characteristics you would like to survey.

   Phrase the characteristics as a question; for example, "How tall are students in my grade?" You will collect data about this question and use the data to produce a graph.

2. Construct a data table.

   Notebook entry: Make sure you design it in a way that will help you record the data you will be collecting. See How To #1, How to Construct a Data Table, for help.

3. Survey the other students in your class and collect data for your question.

   Be sure to show caring and respect for your classmates as they answer your questions honestly.

4. Prepare the graph paper for graphing the data.

   While one person is collecting data, the other should do this.

5. Plot your data on your graph in the form of a bar graph.

   Don't forget to label the graph and give it a title. See How To #2, How to Construct a Graph, for more help.

Wrap Up

Prepare to present your graph to the class. In your presentation you and your partner must answer the following questions:

1. What question did you ask?
2. Why did you choose this question for your investigation?
3. What range does your graph show?
4. How would you describe the diversity your graph shows?
5. Would you say your graph is a normal curve?
6. How well did your strategy work for improving how quickly and quietly you moved into groups?

INVESTIGATION:
A Flag of a Different Color

You probably can remember several times in your life when you had your photograph taken with a flash camera. For a few minutes after the flash, you had a bright spot in front of your eyes that gradually faded. This occurrence is called an **afterimage.** In this investigation, you will explore afterimages and discover more about limits and diversity among your classmates.

Working Environment

Work cooperatively with your partner. One of you should assume the role of Tracker/Communicator and the other should assume the role of Manager. Practice the social skill Stay with your group. Move your desks together to face each other, or sit facing each other at a table. Review the role descriptions to be sure you understand your duties.

Materials

For each team of two students:
- 1 clean, unlined sheet of white paper
- 1 stopwatch or clock with a second hand
- 2 sheets of graph paper

Procedure: Part A—The Social Skill

1. Discuss the importance of staying with your group.

   As you discuss, consider these questions:
   - *How many people would I affect if I did not stay with my group?*
   - *How does it affect my team when I do not stay with my group?*
   - *When is it appropriate to leave my group?*

2. Discuss and record the strategies you will use if one of you does not stay with your group at the appropriate times.

Procedure: Part B—The Flag

1. Construct a data table for this investigation.

   Notebook entry: Construct this in your notebook. Refer to How To #1, How to Construct a Data Table, for help. You will be comparing your data with the rest of the class, so be sure you are doing what you can to promote fair comparisons by having a class discussion to consider the following:
   - *What variables will you control to make sure you all are doing the same thing?*
   - *How can you be sure you are conducting a fair test?*
   - *How can you be sure you will be able to compare your results with your classmates?*

2. Obtain the materials you will need.

   This is the Manager's role.

3. Stare at the bottom right star of the flag in Figure 2.4 for a period of time.

Figure 2.4

Here is a flag of a different color. Use this version of the American flag to test your afterimage. Stare at the bottom right star. Then, fix your gaze on a blank sheet of white paper.

You will take turns doing this. The Tracker/Communicator will do this first.

4. After staring at the bottom right star, immediately shift your gaze to a blank sheet of white paper.

   Have the blank sheet of white paper on your desk next to your book. By immediately fixing your gaze on one spot on the white paper, you should be able to see an afterimage. Try not to blink.

5. Your teammate should take a turn doing steps 3 and 4.

   It is the Manager's turn.

6. After both Team Members have observed an afterimage, repeat this procedure (steps 3 through 5) and record how long the afterimage lasts.

   You will need a method for timing how long the afterimage lasts. Let the Tracker time first, then reverse roles.

7. Repeat step 6 for a total of five times per Team Member.

   Do not take your five turns all in a row. Alternate back and forth with your teammate.

   Notebook entry: Record all times in your data table.

8. Record your data in the class data table.

9. Assist your teacher in constructing a class graph of the afterimage data.

   Remember to show caring and respect for the ideas of your classmates. Think about the following questions and be prepared to answer them.

   - What labels belong on each axis of the graph?
   - What numbers belong on each axis?
   - What would you title the graph?

Wrap Up

Discuss the following questions with your partner and write your answers in your notebook. Be ready to explain your answers during the class discussion.

1. What variables did you control so that you could compare your results with other groups' results?
2. What operational definition did the class use to measure the length of the afterimage?
3. What is the shape of the graph?
4. Explain the pattern the graph makes.
5. What is the class range of afterimage times?
6. Do you think your data suggest that your class is normal? Why or why not?
7. What did you observe when you looked away from the colored flag onto the white paper? Were you surprised?
8. What is a possible explanation for what you saw on the white piece of paper after staring at the flag?
9. During this investigation, how much of the time did you and your teammate stay together: more than half the time, about half the time, or less than half the time?

READING:
The Diversity of Afterimage

What is afterimage? Why did you see a red, white, and blue flag on a white sheet of paper after staring at a green, black, and orange flag? In order to understand this phenomenon, you need to know a few basics about light and color.

White light contains all colors (you can see this if you pass white light through a prism). If you remove one color, let's say green, from white light, then the rest of the colors blend to form the complementary color of green, which is red. This works for all colors. If you remove orange from white light, what remains is the complementary color of orange—blue. The complementary color of black is white. Thus, the complementary colors of green, black, and orange would be red, white, and blue, which are the colors of the flag you saw on the white sheet of paper.

How does the eye process color? Recall from the peripheral vision reading in Chapter 1 that the retina of your eye contains cells called cones, which are responsible for your color vision. The longer you look at a color, the less sensitive to that color the cones in your retina become. Eventually, your eyes become so desensitized to that color that you can no longer see it. When you stare at a green, black, and orange flag for a long time, the cones in

your retina become desensitized to those three colors. When you shift your gaze to a white sheet of paper, your eyes can't see the green, black, or orange components of white light. It's as if those colors have been removed from the white light. What you see are the complementary colors to green, black, and orange: red, white, and blue. It takes time for your cones to become sensitive to green, black, and orange light again. When your cones have regained their sensitivity to these colors, the afterimage fades.

You have explored several ways that people have limits: tracing a star in a mirror, threading an eye bolt with a bolt, and seeing peripherally. Now you know that people's limits for seeing an afterimage depends on how long it takes for the cones in their eyes to resensitize to a certain color. Some people's cones resensitize very quickly, so their afterimages fade quickly. Other people's cones take longer to resensitize, so their afterimages last longer. People are diverse in the time it takes for the cones of their eyes to become sensitive to color again, and that is why afterimage data generate a normal curve.

SIDELIGHT

Blind Spots

Hold your book open at arm's length in front of you and look at the drawing of the cross and the dot.

While closing or covering your left eye, focus on the cross with your right eye and slowly bring the page closer to you. If you do this slowly enough, you will reach a point where the dot becomes invisible. This point is called your blind spot. Both eyes have a blind spot. The procedure used with this drawing of a cross and dot helps you locate the blind spot in your right eye. See if you can locate the blind spot in your left eye using this same cross and dot. How would you have to change the procedure? Are the blind spots in your right and left eyes located in the same place in each eye?

Use this drawing of a cross and dot to determine where your blind spot is.

Just what is a "blind spot"? Inside the back of the eyeball, there is a place in the retina where all the nerves of the eye gather to leave the eye and go to the brain. This spot does not contain rods and cones, so you are blind to any image that falls on this spot. People should show a diversity in how far away their eye has to be from the drawing before the dot disappears. Try it on your friends and see!

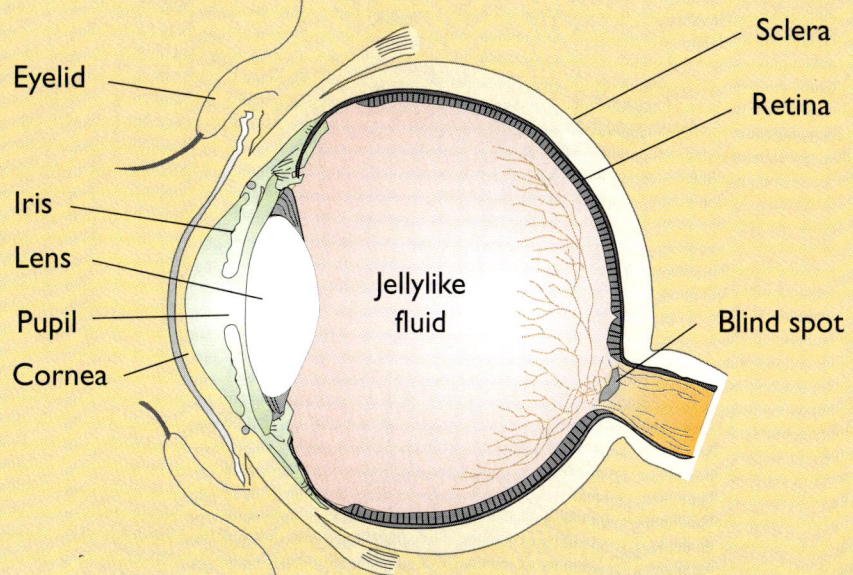

This drawing shows the location of the retina's blind spot in the human eye.

Elaborate

Normal Ranges of Limits and Diversity

CHAPTER
3

Using Limits to Set Standards

In Chapters 1 and 2, you explored human limits and diversity by doing investigations about vision. When you graphed your class data from these investigations, you saw a bell-shaped curve—a normal curve. You also saw that other data you collected from your classmates formed a graph with a normal curve. From these experiences you learned that diversity is normal.

In this chapter you will continue to explore the importance of knowing about the diversity of human limits. You will do this by continuing to investigate diversity in your classroom. This time, however, you will focus on gathering the information you would need in order to set standards when designing a product.

INVESTIGATION:
What Do You Really Know about TV?

Although you may spend a number of hours watching TV every day, how much do you know about how a TV works? In this investigation you will participate in a TV Knowledge Contest. You and your teammates will have a short amount of time to make a list of *everything* you know about TV.

Working Environment

First, you will work cooperatively with your partner. Then you will work as a class. As you work with your partner, make sure you practice your social skill Stay with your group. Your teacher will tell you where your group should sit.

Procedure

1. When your teacher says "Go," your team will have a three-minute brainstorming session to create a list of everything you know about TV.

   Notebook entry: Record your team's list in the Manager's notebook.

2. When your teacher says "Stop," put your writing implement down and hand your team's list to the teacher.

3. Listen and check each item for accuracy as your teacher reads each list to the class.

4. The team with the most correct TV facts on its list wins!

What a great investigation! I wish science could be like this every day. Maybe tomorrow we'll have to tell everything we know about pizza!

Just 30 more seconds, Al. Maybe we could list that soap operas are named soap operas because the sponsors of daytime radiodramas used to be mostly soap companies...put that down.

"We had 32 items on our list and Marie and Al still won the contest!"

"Al watches all those old reruns and he watches MTV. He's a real couch potato. We didn't stand a chance."

"When does he do his homework?"

Wrap Up

Discuss the following questions with your partner. Write your answers in your notebook. Be sure you can explain your answers if your teacher calls on you.

1. How similar were the lists the teams generated?
2. What categories or kinds of information did your classmates list?
3. Is there much diversity in what your classmates know about TV?
4. Do you think your class identified just about everything having to do with TV?
5. Has your team improved its practice of staying together since Chapter 2? If not, discuss strategies that will help you improve.

INVESTIGATION: Taking a Closer Look at TV Pictures

In the last investigation, how many of you listed information about how a television displays pictures? One way to begin exploring the topic of vision and television is to examine TV pictures. Your task will be to observe the television screen carefully and answer some questions about the picture.

Materials

For the entire class:
- televisions
- computer terminals equipped with *TV Teasers* software
- 15 magnifying lenses

Procedure: Part A—Observing a TV Screen

1. Obtain a magnifying lens for your team.

    This is the Manager's role.

2. As a team, visit a station with a television.

    You may work with another team at the station, but the number of students at one station should not exceed four.

3. Carefully observe the television picture.

> ▲ **CAUTION: Prolonged exposure to an operating TV at close range may be harmful. Make any close up observations quickly and do not linger in front of the TV screen.**

4. Record your answers to these questions.

    Be thorough and use your magnifying lens for close up observations. Think about the following questions:

    - *What does the TV screen look like when the power is on and you are getting a picture?*
    - *Step back from the television 3 meters. What happens to the quality of the picture as you move towards the set, beginning at a distance of about 3 meters?*

    Notebook entry: Record your answers in your notebook. Remember to give details.

Procedure: Part B—Observing TV on a Computer

1. Visit a *TV Teasers* computer station.

    You should not begin Part B until you have completed Part A.

2. Complete the first program, Color Pixels.

    There should be no more than six students at any TV Teasers computer station. Work through the first program only. Use the

Working Environment

Work cooperatively with your partner using the roles of Tracker/Communicator, Manager, and Team Member. Work on the skill Stay with your group. You will work at different locations in the room and at your desks for the Wrap Up. Be sure you understand what Stay with your group means when you join other teams at a TV or computer.

materials at the computer station and the instructions in the program.

3. If you are waiting to use a computer, work on the Wrap Up for this investigation while you wait.

Wrap Up

Make sure both members of your team are able to describe your team's observations to the class. After sharing your observations with your partner, write your answers to the following questions in your notebook. You also will share your team's observations with others in your class.

1. What are some of the common observations of a TV screen among the teams in your class?
2. Did other teams observe the same things?
3. List two observations common to all teams in your class.
4. What would you call the tiny elements that compose a TV picture?
5. As a class, decide which teams were best at staying with their group. See whether your teacher agrees. Then choose an appropriate reward for these teams.
6. Using a scale of 1 to 10, 10 being the best, rate your team's use of the unit social skill.

CONNECTIONS:
The Ultimate TV

In your notebook design the ultimate TV. Use your previous knowledge about TV and your recent observations of a TV screen. You can include sketches as well as a description of the ultimate TV. As you continue through the chapter, you will be revising this design as you gather more information about television. Therefore, leave plenty of room in your notebook after this first design so that you can make additional drawings or notes.

READING:
TV Pictures and Color TV

Now that you have looked closely at a TV screen and learned about picture elements in the first program of *TV Teasers,* you are ready to understand more about how a TV works. First, you will learn about black-and-white TV, then you will learn about color TV.

When you looked closely at the TV screen, did you notice that the screen was divided into small sections that appeared as dots? In

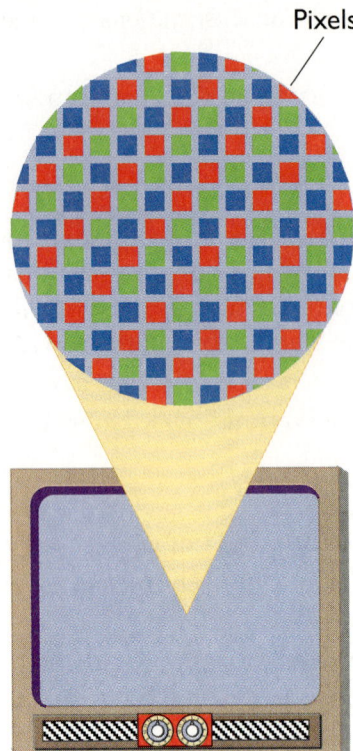

Figure 3.1

This drawing represents a close up view of pixels on a TV screen.

TV Teasers you explored these small sections further and learned that they are called **pixels**. Pixels make up a grid that covers the entire screen (see Figure 3.1).

Each pixel is composed of a material that glows for a brief moment whenever it is bombarded with **electrons.** (Electrons are parts of atoms, and atoms are the tiny particles of which all matter is made.) How do you bombard a pixel with electrons? With an electron gun, of course! Black-and-white TV sets contain an electron gun that fires a beam of electrons at the screen.

An electron gun looks something like the illustration in Figure 3.2. The gun contains a heating element that emits a beam of electrons when it is heated. This electron beam moves quickly across the TV screen from top to bottom, one horizontal row of pixels at a time. The TV camera sends messages to the electron gun that adjust the intensity of the electron beam according to the brightness of the images. As a result, the black-and-white TV picture is composed of many glowing dots of varying shades of black and white.

Color TV works in this same way, with two exceptions. First, a color TV screen is coated on the inside with pixels containing a material called phosphor. The word phosphor comes from the Greek word meaning "light-bearing." Second, color TVs have three electron guns, each one responsible for illuminating a different color on the screen. As you discovered in *TV Teasers*, the three colors of pixels on a color screen are red, green, and blue.

When your color TV receives signals from the TV camera, the three electron guns scan the rows of pixels. Each gun lights up only one specific color of pixels. For example, when the beam from the electron gun specific for red phosphor hits a red phosphor pixel, it produces the color red. The messages from the TV camera indicate which pixel the electron guns should activate on which part of the screen. Just as you would mix different colors of paint to produce new colors, the pixels are close enough together to blend and make different colors.

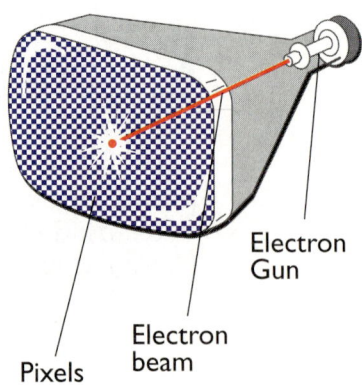

Figure 3.2

If you could remove the outer cabinet of a black-and-white TV, you would see something like the electron gun in this picture.

Figure 3.3

The holes in the shadow mask behind the TV screen guide each electron beam to its specific color of pixels.

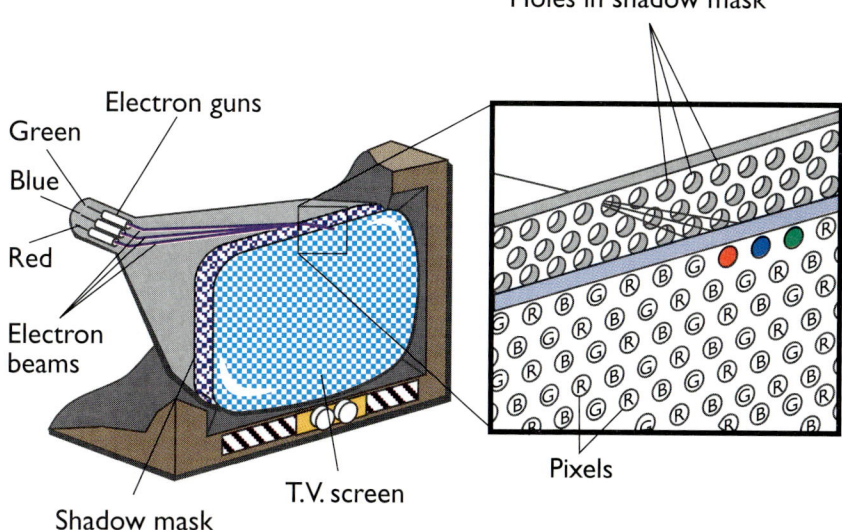

The color TV must somehow keep each electron gun from hitting pixels with phosphor for which it is not specific. For example, the electron beam specific for red phosphor must not strike the pixels coated with blue or green phosphor. To accomplish this, each color TV screen comes with a shadow mask. This mask is full of thousands of holes. Each hole is aligned with a set of red, green, and blue pixels on the screen. The electron beams must pass through the holes in the shadow mask, which focuses the beams on their appropriate pixels, as in Figure 3.3.

SIDELIGHT

Color TV and the Problem of Compatibility

When people develop a new and improved type of technology, it is difficult to introduce it unless it is compatible with the old technology. In the 1950s when companies began producing color TV, two competing systems were available. The CBS system's transmitting signal produced better color in color TV sets, but black-and-white TV sets could not use its signal. The other system's signal, developed by NBC, produced a picture of lesser quality than the CBS's signal did, but black-and-white TV sets could use its signal.

At first, the Federal Communications Commission (FCC) chose the CBS system. Back in those days, however, few people had color TVs. Few people, then, were receiving the programs broadcast in color. Not surprisingly, CBS color programs were canceled after a few months due to the lack of viewers.

Eventually, the FCC switched to the NBC system. Though the picture quality was not as good as it was in the CBS system, all people with a TV set could tune in programs, whether the programs were broadcast in color or in black-and-white.

Explore

INVESTIGATION: A Learning Journey

In this investigation you will embark on a journey around your classroom that will take you to six different locations to explore six different activities. We call these locations learning stations. At each station you will investigate something about the limits of human vision.

At Stations 1 through 4, you will be exploring another limit of human vision. Recall that in Chapter 2, during A Flag of a Different Color, you discovered that people were diverse in their afterimage recovery rates. At Stations 1 through 4, you will explore a phenomenon known as **persistence of vision** in which the human visual system retains the image it sees for a very short time after that image is no longer on the retina. In Station 5 you will explore a limit people have for how they perceive something at a distance. Station 6 is a computer terminal that will enrich your experiences from Stations 1 through 5.

You will circulate from station to station and perform different tasks that will help you set standards for TV pictures. There are six stations and your teacher will assign you to your first station. At each station there will be another team doing the same activity. As you move among stations and work at different stations, use your improving skill of staying with your group. Because you will be generating a lot of different ideas with your partner, be especially aware of your unit skill of showing caring and respect for others and their ideas.

While working at the stations, adhere to these rules:

- Work with your partner at a comfortable pace.

- Stay on task. You will delay a lot of people if you waste too much time at the station.

- Do not rush teams that are ahead of you if they are on task.

- Fill out an evaluation form from the evaluation folder when you finish an activity. When you score your evaluation, you will know if you successfully completed the station. If you did not and you repeat all or parts of the station, fill out a new evaluation form each time you repeat. Put all of the evaluation forms that you fill out in the folder marked "Scored."

- Be completely honest in evaluating yourself. An honest assessment is a good learning tool. A successfully completed evaluation will tell you whether you can proceed to the next station or whether you need to repeat all or parts of the station in order to understand the material.

- Keep the stations orderly. Do not remove any materials. Many students will be passing through the station, so it is important that each team keep the station clean and neat.
- There should be no more than four students at a station.
- Stay with your partner. You are still working as a team.
- Visit Stations 1 through 5 in any order, but do not work through any program at Station 6 until a procedure directs you to do so.

Station 1: Constructing a Spinner

This activity will demonstrate **persistence of vision.** Recall what you read about persistence of vision in the introduction to this activity. Think about the meaning of this activity in terms of television. After the activity try to answer this question: How does your experience with the spinners help explain why the entire TV screen appears lit up even though each electron beam hits only one pixel at a time?

Materials

- 3-by-5-inch index cards, blank
- pencils (not to write with)
- transparent tape

Procedure

1. You will each need one index card and one pencil.

   The Manager should obtain these.

2. Fold the width of the index card in half so you have two rectangles measuring 3 by 2½ inches on each side of the fold.

3. Turn the folded index card so the fold is up and the opening is down.

   The top of each rectangle is now the fold and the bottom of each rectangle is where the card opens.

4. Draw a fish on the front of one of the folded cards and a fish bowl on back of the same folded card. Draw a bird on the front of the other folded card and a bird cage on the back of the same folded card.

   Make sure your drawings are centered from side to side and from top to bottom.

5. Tape the edges of the card together.

   Leave a slit at the bottom.

6. Stick the eraser end of a pencil into the slit at the bottom of the card.

   Push it up until it meets the fold and can't go in any farther.

Working Environment

Work cooperatively in your team of two. You both will be Team Members. One of you also should be the Tracker/Communicator and the other one should be the Manager. The social skill you will be working on is Share your thoughts and ideas.

7. Center the pencil and tape the slit around the pencil closed.

   You should not be able to move the pencil easily from side to side or in and out.

8. Spin the bottom of the pencil back and forth between the palms of your hands so that you see both sides of the card flipping back and forth.

   Spin the pencil slowly at first and gradually increase your speed.

9. Observe what happens as you spin.
10. Trade cards with your teammate and spin each other's pictures.

    If you have the time and the interest, think of some other drawings that you would like to create to spin and observe.

Wrap Up for Station 1

Each person should obtain an assessment form from the evaluation folder and complete it. You will score yourselves and determine whether you may proceed to another station or go to Station 6 and complete the second program, Bird and Cage, in *TV Teasers*.

Record your answers to the following two questions in your notebook.

1. Were you able to see the bird in a cage?
2. Suggest how this experience could explain why the entire TV screen appears lit up even though each electron beam hits one pixel at a time. If you have completed this activity successfully, you may proceed to the second program of *TV Teasers* at Station 6.

Station 2: Continuous Motion through Animation

In this station you will investigate how persistence of vision can make a series of separate pictures seem like continuous motion. As you complete this station, recall the fact that pixels do not keep glowing continuously, yet the entire picture appears lit up.

Materials

- stacks of flip cards 1-39 on card stock
- stapler
- paper
- class data table
- clock or stopwatch

Procedure

1. Make sure you each have flip cards numbered 1 through 39 in order in a stack.

   The stack should have number 1 showing on top and number 39 on the bottom.

Figure 3.4

Before you staple the groups of eight, make sure they are in order.

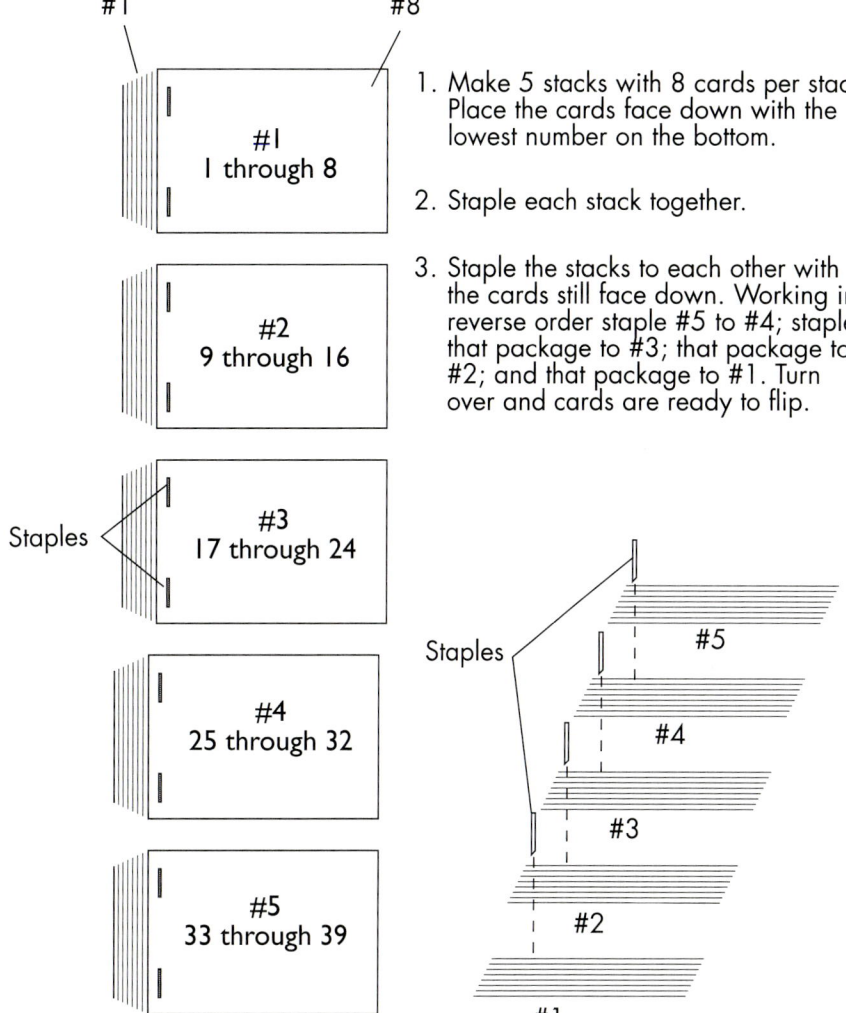

1. Make 5 stacks with 8 cards per stack. Place the cards face down with the lowest number on the bottom.

2. Staple each stack together.

3. Staple the stacks to each other with the cards still face down. Working in reverse order staple #5 to #4; staple that package to #3; that package to #2; and that package to #1. Turn over and cards are ready to flip.

Figure 3.5

Your flip book will look like this. You might have to staple the cards from the top and from underneath the deck to hold them all together.

2. Place the cards in groups of eight so that they overlap and staple each group together. One team member should remove every other card in his or her deck and make two separate decks so that the team has a total of three decks.

   See Figure 3.4. The overlapped edges should not be more than 1 mm wide.

3. Staple all the groups together.

   See Figure 3.5.

4. Flip through the decks to produce a mini-movie.

   Share the deck or decks you made with your partner. Both Team Members should have a chance to experiment with all three decks.

   STOP: Remember, stay with your group means to remain on task with your partner.

5. Determine how fast you have to flip through 39 flip cards for the appearance of continuous motion.

   *This is called the **continuous motion rate** and it is measured in flip cards per second. The simplest way to measure the rate is to have someone first determine (by trial and error) the minimum rate necessary to achieve continuous motion. Then your partner should time how long it takes to flip through all the pages at that rate. You can then calculate the number of flip cards per second as follows:*

   flip cards/sec = total number of flip cards ÷ time in seconds to flip through the cards

6. Record this value in the class data table located at this station.

Wrap Up for Station 2

Each of you should obtain an evaluation worksheet and complete it. Score yourselves and place your worksheet in the folder marked "Scored." You then will need to answer the following two questions in your notebook.

1. When you remove every other frame, do you have to flip the frames at a different rate in order to get continuous motion?
2. Why or why not?

If you have successfully completed this activity, you may proceed to Station 6 to complete the third program, Animation, of *TV Teasers*.

Station 3: Continuous Motion at the Movies

You probably already know that motion pictures, or movies, are a sequence of still pictures. But how fast does the film need to go so that we perceive the separate pictures as continuous motion? You will answer that question as you continue to explore persistence of vision at this station.

Materials

- one 16mm film projector
- 1 reel of 16mm film
- masking tape
- rulers
- stopwatch or clock
- scratch paper

Procedure

1. Use a small piece of masking tape to mark where the film exits the film channel, as in Figure 3.6.

   The projector should be set up so the film is partially wound on the take-up reel.

2. Start the film and note the time or start the stopwatch.
3. Stop the film at the end of 10 seconds.

Figure 3.6

Place your masking tape on the piece of film like this. Make sure you know which edge of the tape is closest to where the film exits the film channel.

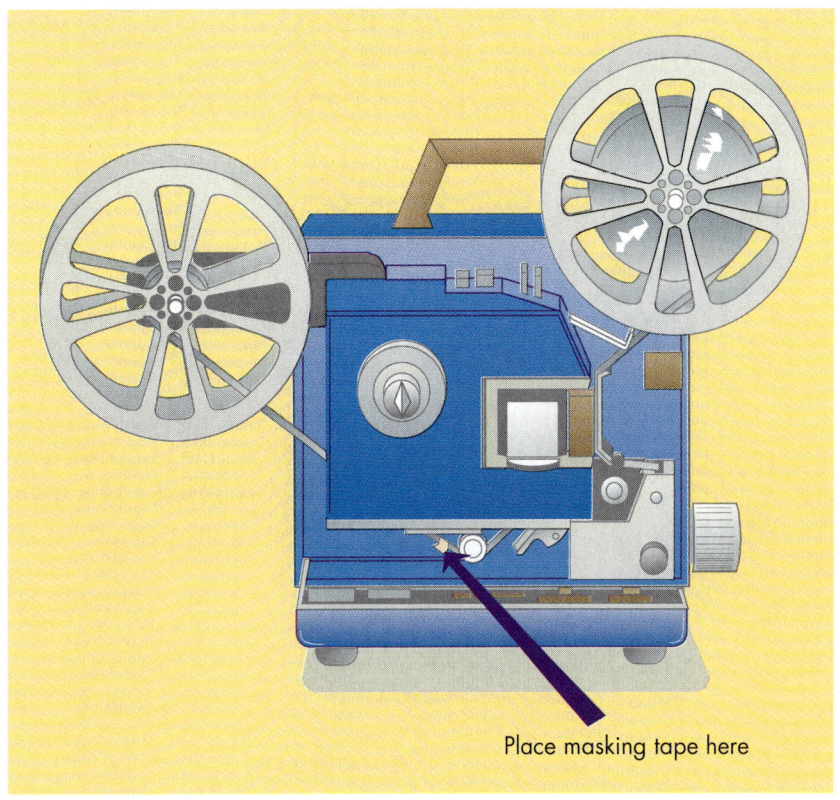

Place masking tape here

The Tracker should say "Stop" at the end of 10 seconds. The Manager should stop the film.

4. Again, mark the new place where the film exits the film channel with a small piece of masking tape.

   You should now have a strip of film marked on both ends with a small piece of masking tape.

5. Measure the length of film that went through the projector in 10 seconds.

   You will be measuring the length between the edges of the two pieces of tape that were closest to where the film exited the film channel. Make your measurements in centimeters.

6. Calculate the length of film that goes through a projector in one second. Find this value in centimeters per second.

   Record this value in your notebook. Remember to show caring and respect for others and their ideas.

7. Determine how many frames of film move through the projector each second and record this value.

   Use the length of film between the two pieces of masking tape and your previous calculation from step 6.

   For example:

   Number of frames of film in the length of film from step 6 = "x" frames

   Film rate = "x" frames per second

Explore — Using Limits to Set Standards

8. Rewind and replace the film on the receiving reel as you found it.

Remember to carefully remove the pieces of masking tape from the film.

Wrap Up for Station 3

Each person on your team should obtain an evaluation worksheet and complete it. Score yourselves and put your evaluation sheet in the "Scored" folder. When you have completed the activity successfully, you may proceed to any other station except Station 6.

Station 4: Flicker-Fusion Frequency

Why do we see a whole picture on a TV screen instead of dots glowing one by one? In this station you will find the answer to that question by exploring a phenomenon called **flicker-fusion**.

Materials

- 1 variable strobe light
- paper
- rulers
- class data table

Procedure

1. Prepare a data table for your team.
2. Start the strobe light at 15 flashes per second.

   You will find the strobe light turned off and set at 15 flashes per second. When you finish using the strobe light, leave it in the same condition as you found it.

   ▲ **CAUTION: Never stare directly at the strobe light. Also it is dangerous for humans to see flashes below 15 flashes per second.**

3. Stand to the side of the strobe light and move your arm up and down slowly.

   The Tracker should do this first.

4. Observe what your teammate's arm appears to be doing and record your observations in your data table.

   The Manager should observe and record first. The Tracker should turn off the strobe light when you are not using it. Strobe lights burn out quickly.

5. Continue to make these observations and record them as you increase the flash rate on the strobe light.

   The Tracker should increase the flash rate in increments of five (20 flashes per second, 25 flashes per second, 30 flashes per second, 35, and so on) up to 60 flashes per second. The Tracker should continue to

Figure 3.7

Al and Ros are demonstrating the proper positioning for watching your partner in this activity. Be sure that you do not look directly at the strobe light. Notice how Al looks at Ros from the side of the strobe light, while Ros keeps her back toward the strobe light.

move his or her arm in the strobe light path each time. The Manager should record his or her observations each time.

STOP: Are you staying with your group?

6. Trade places with your teammate and repeat steps 2 through 5.
7. Determine from your data tables the frequency at which you no longer perceive a flicker.

   *This is called your **flicker-fusion frequency.***
8. Record the flicker-fusion frequency for both members of your team.

   Record these on the class data table at this station.

Wrap Up for Station 4

Obtain evaluation sheets for both team members and complete them individually. Score yourselves and place your evaluation sheets in the folder marked "Scored." When you have successfully completed the activity, you may proceed to any other station or go to Station 6 and complete the fourth program, Color Mix by Flashing, of *TV Teasers*.

Station 5: How Many Lines Per TV Picture?

At this station you will explore another vision factor that influences the design of a television. You know that a TV picture is made of horizontal rows of pixels. But a TV picture doesn't look like a bunch of dots unless you examine the screen up close. Just how far away from a TV set do you need to be so that you see a complete picture, rather than separate dots? How do different colored pixels blend to produce a screen that displays all colors? Explore and find out.

Materials

- a propped up page of BLM 3.13, Black Lines
- meter sticks or tape measures
- class data table
- masking tape
- rulers
- calculators
- paper

Procedure

1. Stand directly facing the page of black lines.

   Take turns doing steps 1 through 3. The Tracker should do this first.

2. Walk backward slowly just until you can no longer see the spaces in between the lines and can no longer make out the distinct lines. Stop at this point.

   As you walk backwards, your teammate should make sure you don't trip or bump into anything.

3. Measure this distance.

   Have your teammate lay a strip of masking tape from the wall to the heel of your foot and then measure the length of the masking tape (in meters).

4. Record this value in your notebook.

5. Trade places with your teammate and repeat steps 1 through 3.

6. Now it is time to calculate each team member's "D to s" ratio or **D/s ratio.**

   These calculations will go more smoothly if you show your teammate that you respect his or her ideas.

   - First, measure the space between the lines on the lined paper and confirm that the spacing is 2 mm. Because "s" stands for the space between the lines, $s = 2$ mm.

- Next, convert into millimeters (mm) the distance you stood from the lined page when the lines were no longer distinct. Do that by multiplying the distance (which should be in meters) by 1000. "D" stands for the distance you stood from the lined page when the lines first disappeared. If a person were 3.6 meters away from the lined paper when the lines were no longer distinct, then D = 3.6 X 1000 = 3,600 mm.
- Now, calculate the ratio of D to s. Do this by dividing the D by s.

$$D/s = \frac{3{,}600 \text{ mm}}{2 \text{ mm}} = 3{,}600 \text{ mm} \div 2 \text{ mm} = 1{,}800$$

7. Record each person's D/s ratio in the class data table at this station.

Wrap Up for Station 5

Answer the following questions as a team. Record one set of answers for your team using either the paper at the station or your notebooks.

1. If someone could see small details at a great distance, would his or her D/s ratio be small or large? Why?
2. Suppose you had a D/s ratio of 100. If a pair of lines were drawn 2 mm apart, how far away would you have to be before you could no longer see separate lines?
3. Assume the lines you observed on the lined paper were rows of pixels on a TV screen. If you had a TV screen that was 30 cm (300 mm) high, how many lines would there be if the lines were separated by 1 mm?
4. About how far away from the TV in question 3 would you have to sit to see a clear picture? (Use your own D/s ratio to calculate this.)
5. By now you should have noticed that the number of lines of pixels you need in order to have a clear TV picture depends on how far away you plan to sit. How does the distance you sit from the TV set affect the number of lines that you need?

Obtain an evaluation worksheet for each of you and complete them. Score them against the key and place in the folder marked "Scored."

Station 6: *TV Teasers* Computer Station

You may visit this station and complete a program *only* when you are specifically told to do so in Stations 1 through 5.

Materials

- computer with color monitor
- *TV Teasers* software program

Procedure

The instructions and procedures for this station are contained entirely in the software program.

Wrap Up for All Stations

Part A

When every team has completed all six stations, work with your partner at your desks to complete this Wrap Up. Record answers to questions in your notebooks and be prepared to explain your answers in a class discussion.

Your teacher will assign each team a data table from either Station 2, Station 4, or Station 5 to graph.

- With your partner, use your assigned data table to construct one graph of data from your entire class. If you encounter any difficulties constructing your graph, have the Communicator get help from other teams using the same data table.

- When you have completed your graph, consult the other teams who constructed the same graphs to see whether they are similar.

- Save your class graphs in the Tracker/Communicator's notebook until you have completed the connections, Your Experiences at the Stations.

Part B

Compare your graphs with other teams' graphs. Make sure you see what shape of graphs the other data produced.

1. At which stations did you see the most class diversity?
2. At which stations was it difficult to measure diversity? Why?
3. Which of the class graphs is the best representation of a normal curve in your class population?
4. Draw a normal curve on a set of axes. Put an X on the curve where you feel your team would fall when compared with others in your class in using the unit skill. Put an ◯ where your team would fall in staying with your group. Explain your reasons for the placements you chose.

CONNECTIONS: Your Experiences at the Stations

Work individually to answer the following. Write your solutions in your notebook and be ready to share your answers in a class discussion.

1. If pixels on an operating TV screen do not glow continuously, how does the entire picture appear constantly lit up?
2. Turn to the place in your notebook where you drew your ultimate TV design in the connections, The Ultimate TV. Review your design and make a list of the things you would now consider when designing your own TV. Record your list in your notebook.
3. Share your list during the class discussion and add to your list any new ideas you hear.

INVESTIGATION:
The Optimal TV Viewing Distance

In your journey through the stations, you learned about a variety of diverse human limits that affect TV design such as persistence of vision, flicker-fusion frequency, and viewing distance ratios (D/s). If you were going to design your own TV screen, you would have to consider the diversity in people's preferred viewing distances. That way you could manipulate factors like lines per screen in order to make the best TV picture possible. But what viewing distance should you assume? Should you use yourself as the example? Should you determine the average viewing distance based on the average size of a den or living room? These are not the most scientific ways to go about deciding the optimal TV viewing distance. A better way is to collect data—something that you are getting good at! In this investigation you will collect data that will help you determine the average TV viewing distance of a group of people who are the best TV critics in the world: you and your classmates.

Working Environment

Work cooperatively as a Team Member with your partner. You also will need the roles of Communicator and Tracker for this investigation. Push your desks together or sit side by side at a table. Your teacher will tell you how to go about conducting a survey for this investigation. Work on the social skill Move into your groups quickly and quietly.

Materials

For each team of two students:
- graph paper
- calculator
- scratch paper
- meter stick or tape measure

Procedure

1. Construct a data table for this investigation.

   You will need one per team. The Tracker should do this. Remember to read through the entire procedure first before you attempt to construct your data table.

2. Conduct a survey of all the students in class to find out how far they like to sit from the TV while viewing a program.

   Recall how teams went about conducting the last survey in Chapter 2.

Explore   Using Limits to Set Standards

3. Record your data in your data table.

4. Use any, all, or none of the materials your teacher provides to determine the average TV viewing distance of the students in your class.

Wrap Up

With your partner, complete the following tasks in your notebook.

1. Prepare a summary of your method and results to present to the rest of the class. Include in your summary all of the following:

   - your survey method,
   - your range,
   - your average value, and
   - a comment on the amount of diversity you observed among your classmates.

2. Rate your team on its ability to move into groups. Choose either (a) couldn't be better, (b) needs a little improvement, or (c) couldn't be worse. Explain your choice.

READING: Setting Standards and Human Factors

In the previous investigation, you considered the limits of people when determining the optimal TV viewing distance. Why would anyone need such information? Recall from Chapter 2 the jeans company that used information about the heights of twelve-year-olds to manufacture jeans that fit.

In setting standards like sizes, we need information about the way people *differ*. These differences are called **human factors.** Human factors are very important to consider when designing any product. Manufacturers want to be sure that their products fit the people who will be using them.

If you are designing clothes, it is simple to see what we mean by having the product fit the user. But what does it mean to have a skateboard, a bicycle, a calculator, or even a TV "fit" the person who uses it?

Take a calculator, for example. A calculator has to be the right size for people to hold in their hands, but not all people have the same size hands! When designing a calculator, one also has to make sure that the keys are far enough apart so that most people can touch them without hitting two keys at once. And what about left-handed people? The display needs to be small enough to fit on the calculator but large enough for people to see when it is held at a reasonable distance from their eyes.

Things like hand size, finger size, and limits of vision are called human factors. You must consider human factors when you are designing a product that humans will use so that you come up with a standard product or one that fits the most people. It is important to make sure that products are safe, comfortable, and easy to use.

Stop and Discuss

You have spent a good deal of this chapter discovering some human factors that influence the way people design TVs.

1. List four of these factors.
2. Describe any limits you think these human factors will have on the design of your ultimate TV.
3. Can you design a standard TV that suits most people?

CONNECTIONS: Clear As a Belle

Work with your partner to read the following discussion and then plan and present the activity. Be sure to show caring and respect for each other's ideas.

Imagine that you and your partner began work at an innovative company called Belle Television and Electronics. At Belle, the employees not only design and manufacture TVs, but they also create advertisements to sell their TVs. You and your partner, recent college graduates, were perfect candidates for jobs at Belle because you each received degrees in engineering and advertising. Each employee at Belle is hired with the understanding that his or her employment is probationary until he or she can design a new TV and create and demonstrate a television commercial for it. Before you set out to accomplish your task, you had several meetings with the board of directors and the senior partners of Belle. At these meetings, the board made several decisions that you and your partner must adhere to. The decisions are as follows.

Meeting 1: You and your partner will work cooperatively with one of you acting as Manager and the other acting as Communicator.

Meeting 2: You will record in your notebook anything you do during your task.

Meeting 3: You will set standards for the design of your TV before you send it to the assembly line for manufacturing. The senior engineer described a standard this way:

> "A standard is a set of guidelines that industries use to maintain product consistency. Standards help us make products we can depend on. Imagine if we had no standards for the sizes of light bulb sockets and light bulb bases."

The senior art executive jumped in and said:

> "Standards are guidelines we set for the safety and well-being of society. Imagine if there were no standards for how fast cars could travel."

The president of Belle had the final word (naturally):

> "The bottom line for these two new employees is that standards are guidelines we use to design the best product. We need you two rookies to set standards for the best TV picture. We want to beat all the other companies in TV sales."

Meeting 4: The standards you set will be the same for all TVs regardless of screen size.

Meeting 5: You will concentrate on setting standards for the following three television components as you design your TV:

1. How many pictures per second?

    In order to decide on an answer to this question, you might review your findings from Stations 1, 2, and 3 on continuous motion and from Station 4 on flicker-fusion frequency. Also review the information about these phenomena. The goal is to produce smooth continuous motion that is pleasing to the viewer. After you arrive at an answer, justify it with a written list of reasons.

2. How many horizontal lines of pixels?

    You might review your findings from Station 5 that dealt with lines per screen. How will your understanding of D/s ratios influence your answer to this question? You also might consider the results of the survey about people's TV viewing distances.

3. How many pixels per horizontal line?

    Your answer to this probably will depend on whether the shape of your TV is rectangular or square.

Meeting 6: You are free to be creative in producing your TV commercial as long as you use the company slogan, which the company has been using for 25 years: "With a Belle TV, you can count on a picture that's as clear as a Belle."

You and your partner are now ready to begin your task: Design a TV with a clear picture by considering the components described in Meeting 5 and produce and present a TV commercial to sell your TV. The chairman of the board, affectionately nicknamed "Teacher" at Belle, will tell you how much time you will have and what materials you may use to design your TV. This person also will tell you when you will present your commercial.

READING:
The National Television Systems Committee Uses Human Factors

In the previous connections section, you considered the way a TV works and certain limits to human vision that influence the design

SIDELIGHT

Interlacing

In 1940 the National Television Systems Committee (NTSC) decided that one of our TV standards would be that 30 pictures per second should appear on the screen. As you learned earlier, this rate would result in the perception of continuous motion. Unfortunately, at this rate, the light did not look continuous and people saw an annoying flicker. Researchers discovered that the light would have to flash about 45 times per second before the flicker would disappear.

The NTSC had a problem. They wanted to limit the number of pictures per second to 30 and still eliminate the flicker. Fortunately, engineers came up with a technique to avoid sending more pictures than necessary and still get rid of the flicker. This ingenious technique is called "interlacing." In interlaced television, the electron gun first traces or activates every other line of pixels from the top of the screen to the bottom. The electron gun then returns to the top, and traces all the remaining lines of pixels. In other words the electron gun traces half the picture every $1/60$ of a second or a complete picture every $1/30$ of a second. A complete picture, then, requires two tracings over the picture tube from left to right and top to bottom. There is no flicker because the eye cannot detect flicker in such a small area (another human limit). Even though 30 complete pictures appear every second, our brain perceives 60 complete pictures per second.

---

of TVs. The National Television Systems Committee (NTSC) is the organization that considers those same components in setting the standards we currently use to produce TV pictures in the United States and Japan. The standards are called the **NTSC standards.** It is indeed amazing that standards established more than 50 years ago are still in use today. The connections section you just completed is similar to what the NTSC did in 1941.

One of the reasons the standards have stayed the same for so long is that the NTSC based them on human factors. The NTSC had to answer the same questions that you answered. In order to set the standard for how many pictures per second should appear on a TV screen, they used what they knew about motion pictures and the research on persistence of vision and flicker-fusion frequency and arrived at 30 pictures per second. In order to get rid of the flicker (remember, humans stop seeing flicker at about 40 to 50 flashes per second), they used a clever technique called interlacing.

In order to set standards for the number of lines per picture, they used information about how well people see at a distance. They used a D/s ratio of 2000 and assumed that people would sit about 15 meters (4 feet) away from a 30 cm (1 foot) high TV screen. This information helped the committee decide on a standard of 525 lines per screen.

TV screens can be shaped as circles, squares, or rectangles. So the committee had a difficult time deciding what shape the screen should be. Because most of the early programs on TV were movies, they finally decided on a rectangular screen that was about the same shape as a movie screen.

Using a rectangular screen, the NTSC determined how many pixels per line were necessary. They used a ratio of 4 to 3 for the height-to-width ratio of a TV screen, which gave 700 pixels per line (or $4/3$ of 525). After much discussion and investigation, they decided that 700 pixels per line was more than necessary. To keep the TV equipment as simple as possible, the committee decided that 630 pixels per line was enough.

Your team probably developed standards that are different from those developed by the NTSC, and that's okay. The important thing is that you researched human factors, including human limits, and explored how you can use those two things to set standards. This will be an important skill to carry with you into the next chapter.

SIDELIGHT

TV Today and Tomorrow

A new type of TV set called digital or Improved Definition TV (IDTV) is now available for consumers to purchase. IDTV is called digital TV because the TV signals are digitized (a continuous signal coded as a 1 or a 0). Unlike regular TV, in which the signal from only one station can be selected at a given time, IDTV can digitize the signals from many stations and store them in its computer memory.

Because video information is stored in the TV's computer memory, you can choose to watch one station at a time or view multiple stations at the same time. For example, a viewer can watch a movie on half of the screen and a football game on the other half of the screen. The audio part of the TV set, however, can provide sound for only one program at a time.

Another advantage of digital TV is that the TV picture no longer has to be interlaced. The computer uses an averaging technique and fills in the missing lines. This results in an improved picture. The viewer watching digital TV actually sees 60 complete pictures per second instead of 60 half-pictures per second.

Currently, research and development laboratories in Japan, the United States, and several European countries are experimenting with new systems for broadcasting TV pictures that are of much higher quality. These systems are called High Definition TV (HDTV) because they will have about twice as many lines per picture as conventional TVs. The term high definition means greater detail.

The Japanese already are testing a HDTV system that uses a satellite to broadcast its signals. The main drawback of the Japanese system is that regular TV sets cannot receive the HDTV signal. In 1989 our Federal Communications Commission (FCC) decided that future HDTV systems in the United States must send signals that both regular TVs and new HDTVs can receive. Many companies are now trying to design and develop HDTV systems that are compatible with regular TV systems.

CHAPTER 4

Using Diversity to Set Standards

Take a ride along your local roads and you will find a diversity of drivers: drivers who weave back and forth across their lanes, drivers who crawl along at a slow pace, and drivers who gun their engines at the stop light and leave you in a cloud of dust. In previous chapters you investigated diversity in human limits and how we can use diversity and limits to set standards for televisions. Televisions aren't the only things that require standards in our society. In this chapter, you will consider how our knowledge about human diversity can be useful in setting standards for driving.

 **INVESTIGATION:**
Watching *The Final Factor*

You might have brothers or sisters who recently have obtained a driver's license. You also might be anticipating the day when you will be eligible for your driver's license. What does it take to learn how to drive safely? This investigation will introduce you to some of the important factors involved in driving.

Materials

For the entire class:
- 1 copy of the video *The Final Factor*
- 1 television set with VCR

Procedure

Watch the video *The Final Factor*.

You might want to take notes if that helps you remember what you watched. Pay close attention to the actions and reactions of the drivers.

Wrap Up

With your partner discuss the following questions and write the answers in your notebook. Prepare to share your answers with the class.

1. Describe whether or not you think the word "factor" is used the same way in the video as it was in Chapter 3 in the phrase "human factor."

2. What are some factors that influence how fast people should drive when operating an automobile?

3. Which of these are human factors?
4. Decide and record what the speed limits for automobiles should be in the following areas. Your limits don't have to agree with the legal speed limits for these areas, but you must be able to explain your reasons for your choices.
   - school zone
   - your neighborhood
   - divided highway (interstate)
   - rural highway

 **INVESTIGATION:**
Green Light—Red Light!

Did you ever play the game "Green Light—Red Light"? One person, the caller, turns his or her back on a line of players and yells, "Green Light!" That's the signal for the players to charge as fast as they can toward the caller. Suddenly, without warning, the caller yells, "Red Light!" and turns to face the charging players. The players then try to stop as quickly as they can. If the caller catches anyone still in motion, that person must go back to the starting position. This simple game can teach you about some important factors at work when you are driving a car. The next investigation will help you relive this childhood game and learn something new about human factors.

Working Environment

You will work individually and as a class. Your teacher will provide a large space in which many students can run at the same time and will designate some of the students to be class recorders. Others will run in the races.

Materials

For the entire class:
- a large area for running
- 4 meter sticks or tape measures
- 1 roll of masking tape
- 10 sheets of plain white paper
- 10 pencils

For each student:
- comfortable clothing
- shoes suitable for running

Procedure: Part A—Green Light

1. As a class, decide on a starting line and a finish line.
   You might mark these with masking tape, a flag, or people.
2. Assemble in a long line as if you were getting ready to race one another.
   Your teacher and class recorders will stand at the place that marks the finish line. Make sure you know where that finish line is.
3. When your teacher says, "Green Light!" run to the finish line.

4. As soon as you reach the finish line, stop running. Freeze at the point where you come to a complete stop.
5. Record the names of the first, second, and third place finishers.

    The recorders will do this. Listen as your teacher tells you who the first, second, and third place finishers are.

6. Measure the distance from the finish line to the heels of the three people who stopped farthest from the finish line.

    Your teacher will do this or assign a recorder to do this. The recorders will record these names and distances.

Procedure: Part B—Red Light

1. Return to the starting line and line up again.

    Your teacher and class recorders will now stand aside so you can't see where the finish line is. If you marked a finish line, remove the marker.

2. Start running when your teacher calls, "Green Light!"
3. Stop running when your teacher calls, "Red Light!"

    Come to a complete stop and stay there. Your teacher will now measure the distance from the starting line to the heels of the three students who stopped closest to the starting line. Do not move from your place once you have stopped. The recorders will record the names of these three people.

Procedure: Part C—Green Light and Red Light

1. Repeat the Green Light—Red Light game you played in Part B.

   This time see how quickly you can stop when your teacher calls out, "Red Light!"

2. Measure the distances from the starting line to the heels of the three students who stopped closest to the starting line.

   Your teacher will do this or will assign a recorder to do this. Remember not to move from your stopping point.

Wrap Up

Your teacher will have the recorders share their lists of names with the rest of the class. While answering the following questions, refer to these names. Write your answers in your notebook and prepare to share your answers with the rest of the class.

1. Were the people who came in first, second, and third in the race in Part A the same people who were the farthest from the finish line?
2. Explain why it makes sense that the people who were the first, second, and third place finishers were the same as the people who were the farthest from the finish line.
3. In Part B what did the three people who were closest to the starting line do differently from the other students?
4. In Part C how did people change their running style in order to be the first to come to a complete stop?
5. What types of diversity did you observe when you played Green Light—Red Light?
6. What factors might account for the diversity you observed?

READING: The Three Phases of Stopping

Why do different speed limits exist for different parts of the city and country? For example, why aren't people allowed to drive as fast on a residential street as they are on the open highway? The answers to these questions are related to the limits of humans and their abilities to stop their vehicles during a sudden emergency. Stopping is not an instantaneous event. Stopping takes time, no matter how quickly it seems to occur.

There are three phases in the stopping process. Recall the events that took place in the investigation Green Light—Red Light. In Part B you raced with your classmates until the caller yelled, "Red Light!" Although you might not realize it, first your ears received the sound and then your brain perceived the words. The time that elapsed between hearing and perceiving is called the

perception time. During your perception time, you continued to run before you even began to stop. The distance you traveled during your perception time is called the **perception distance.** This is the first phase in the stopping process. Everyone takes some time to notice or perceive an event. This is especially true if the event is a surprise. Not every person's perception time is the same. You might have been slightly faster or slightly slower than your classmates at perceiving the "Red Light" signal. Imagine how difficult it would be to determine the average perception time for all five billion humans. Instead scientists have used a normal curve to determine the average perception time for all humans—0.75 seconds.

After your brain perceived the words "Red Light," it took a little more time for you to react to the words and begin to stop yourself. The time it took for you to react is called the **reaction time.** Like perception time, reaction time varies from one person to another. The distance you traveled during that time before you began to stop is called the **reaction distance.** This is the second phase of the stopping process.

Once you "applied your brakes," it took you a certain amount of time to come to a stop. The time that passed between when you first tried to stop and when you finally came to a complete stop is called the **skidding time.** During the skidding time, you traveled a distance known as the **skidding distance.** The skidding distance is the third and final phase of the stopping process.

If you add together the perception time, the reaction time, and the skidding time, you will have the total amount of time that it took someone to come to a complete stop. Similarly, if you add

Figure 4.1

What human factor might have affected the stopping process that resulted in this multiple car accident?

together the perception distance, the reaction distance, and the skidding distance, you will have the total distance someone covered before coming to a complete stop.

In the investigation Green Light—Red Light, you probably realized that you could affect your total stopping distance by changing your speed. It probably took you longer to come to a complete stop in Part B than it did in Part A. In Part A you knew in advance where you were going to stop. This means that your perception time wasn't as long, and therefore it didn't add as much distance to your total stopping distance. This wasn't the case in Part B. You also might have realized that the faster you go, the longer your skidding distance will be. Therefore, by the time you raced in Part C, you probably reduced your speed and consequently your total stopping distance.

Now consider the following example of how the phases of stopping might apply to real life.

> You are riding down the highway in your family car. As the car roars along at 60 miles per hour (100 km per hour), the driver suddenly sees something in the road ahead. "What's that?" the driver asks. You both realize what it is at the same time. It is a truck stalled in your lane. The driver slams on the brakes and your car skids to a stop just inches from the back of the truck.
>
> After breathing a sigh of relief that you didn't hit the truck and that you both were wearing your safety belts, you say, "Wow! You barely had enough total stopping distance to avoid hitting this truck! You must have incredible perception and reaction times!"
>
> "Huh?" asks the dazed driver. "What are you talking about?"
>
> Happy to demonstrate your scientific know-how, you say, "You see, back there when you saw something in the road, the car kept going while you tried to figure out what it was. That was your perception distance. Then, in the time it took you to take your foot off the gas pedal and hit the brake pedal, the car kept moving. This distance is called your reaction distance. Finally, after you slammed on the brake, the car skidded to a stop. The distance the car moved while it skidded was your skidding distance. Your perception distance, reaction distance, and skidding distance all combine to make your total stopping distance. Either you have quick perception and reaction times, or you just barely had enough space to stop for the speed you were traveling!"
>
> You turn to look at the driver, who has fainted at the wheel. "Oh well," you say. "I wish I knew all of these distances for the speed we were going. There's no way that I could measure them now. Maybe I'll learn about that in science class tomorrow."

In the following activities, you will explore the relationships among perception distance, reaction distance, and skidding distance at different speeds.

CONNECTIONS: How You Can Affect the Three Phases of Stopping

Discuss your answers to the following questions with the rest of your class.

1. In Part A of Green Light—Red Light, you might have noticed a diversity of perception times among your classmates. What factor reduced your perception time when you reached the finish line?

2. Can you recall any event in the video *The Final Factor* where someone adversely affected his or her perception time—that means an event where someone did something to make his or her perception time longer?

3. When you ride in a car, what do you see drivers do that would adversely affect their perception times?

4. Can you think of any instances in *The Final Factor* in which someone adversely affected his or her reaction time?

5. When you ride in a car, what do you see drivers do that would adversely affect their reaction times?

6. Describe an example from this video in which someone's skidding distance was increased.

7. Think of the scenarios presented in the video. What things do people do that dangerously limit (or decrease) the space in which they have to stop?

INVESTIGATION: Your Personal Reaction Time

You have seen that the faster you go, the longer it takes you to stop. Before you can stop, however, you must first react. Once you know your reaction time, you can determine reaction distances for different speeds. But how long does it take you to react to a surprise event? You will know the answer to this question by the time you finish this investigation.

Materials

For each team of two students:
- 1 meter stick
- graph paper
- 1 ruler

Working Environment

Work cooperatively in your team of two. Besides being Team Members, you will need a Communicator and a Manager/Tracker. Work on the social skill Stay with your group. Push your desks together to face each other, or sit facing each other at a table. Create a space beside your work area in which you both can stand.

Procedure

1. Hold the meter stick as shown in Figure 4.2.

   The Communicator should do this, holding the meter stick so that the number 1 is at the bottom.

2. Hold your fingers near the meter stick at the opposite end (see Figure 4.3).

   The Manager holds his or her fingers near the bottom of the meter stick without touching the stick. He or she will try to catch the meter stick when the Communicator drops it.

3. Catch the meter stick.

   Communicator: Let go of the meter stick without any warning. Because you want to measure reaction time, it is important that you give no warning.

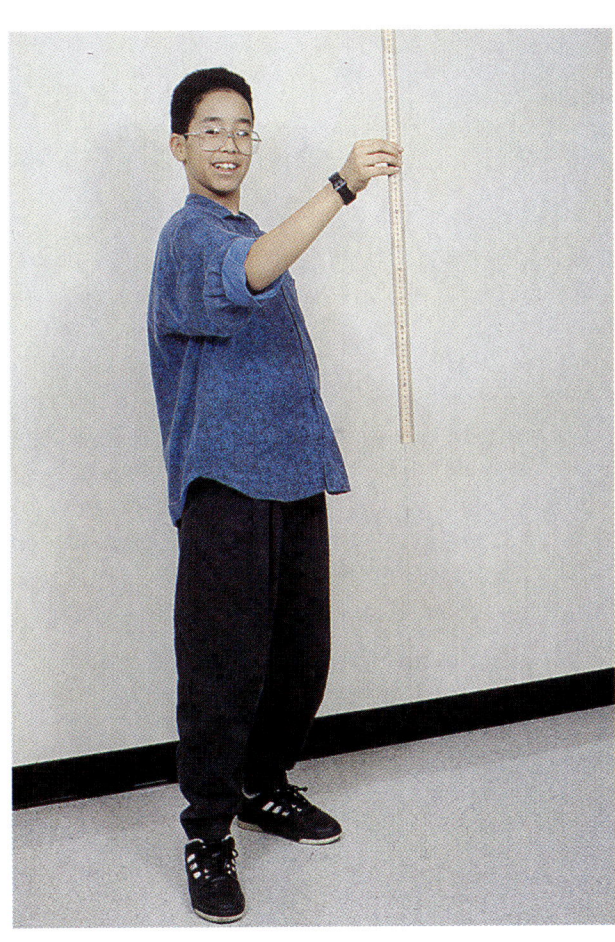

Figure 4.2

Hold the meter stick in the middle with the number 1 at the bottom.

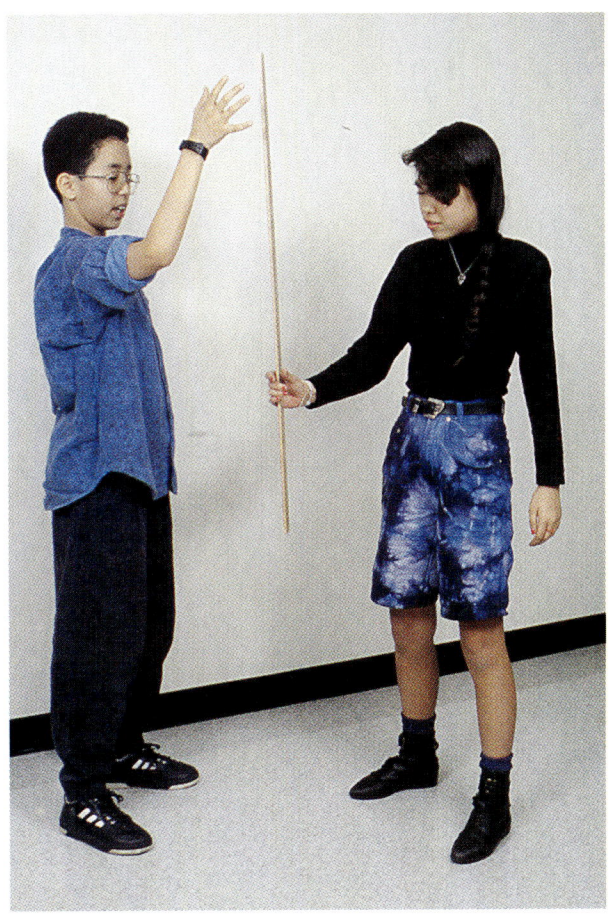

Figure 4.3

The Manager should catch the meter stick when the Communicator drops it.

Elaborate

4. Measure the distance between the bottom end of the meter stick and the point where the Manager caught the meter stick.

   The Communicator should measure this distance in meters as shown in Figure 4.4 (10 cm = 0.10 m, 27 cm = 0.27 m, 42 cm = 0.42 m, and so on.) Record the distance in the Manager's notebook.

5. Repeat steps 1 through 4 until each of you have had at least three tries catching the meter stick.

   Take turns dropping the meter stick for each other. You are finished when each of you has recorded three distances in your notebook.

6. Average your three distances.

   Record this average distance in your notebook. We will call this distance the falling distance.

7. Use the graph in Figure 4.5 to determine your average reaction time.

   Look at the example on the graph. Notice that if your average falling distance is 0.60 m (remember that is 60 cm), then your reaction time is 0.35 seconds.

8. Record both your team member's and your own reaction times on the class data table your teacher has prepared.

   This is the time that you have determined from the graph in Figure 4.5.

9. As a team, graph the class reaction times.

   Notebook entry: Each of you should have this graph in your notebook. If you need to review graphing skills, refer to How To #2, How to Construct a Graph.

Wrap Up

You will need to look at both graphs, the one in Figure 4.5 and the one you made of your classmates' reaction times, to answer these questions. Discuss your answers to the following questions with your partner. Record your answers in your notebook. Be ready to explain your answers when your teacher calls on you.

1. Why was it important to average your three falling distances?
2. Why do we use a graph to determine reaction times? Why not just have someone time how long it takes for you to catch a meter stick?

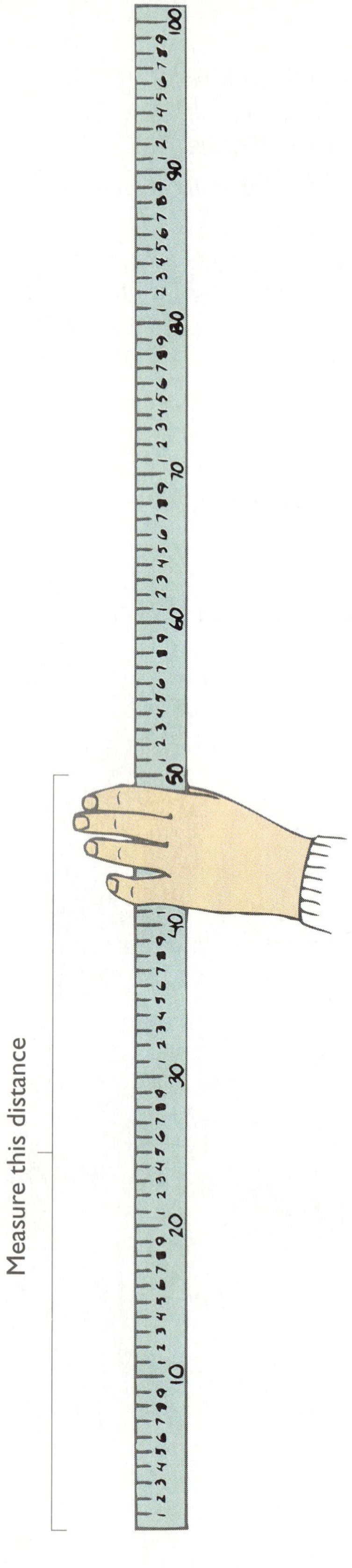

Figure 4.4

Measure the distance from the first mark at the bottom to the top finger where you caught the stick.

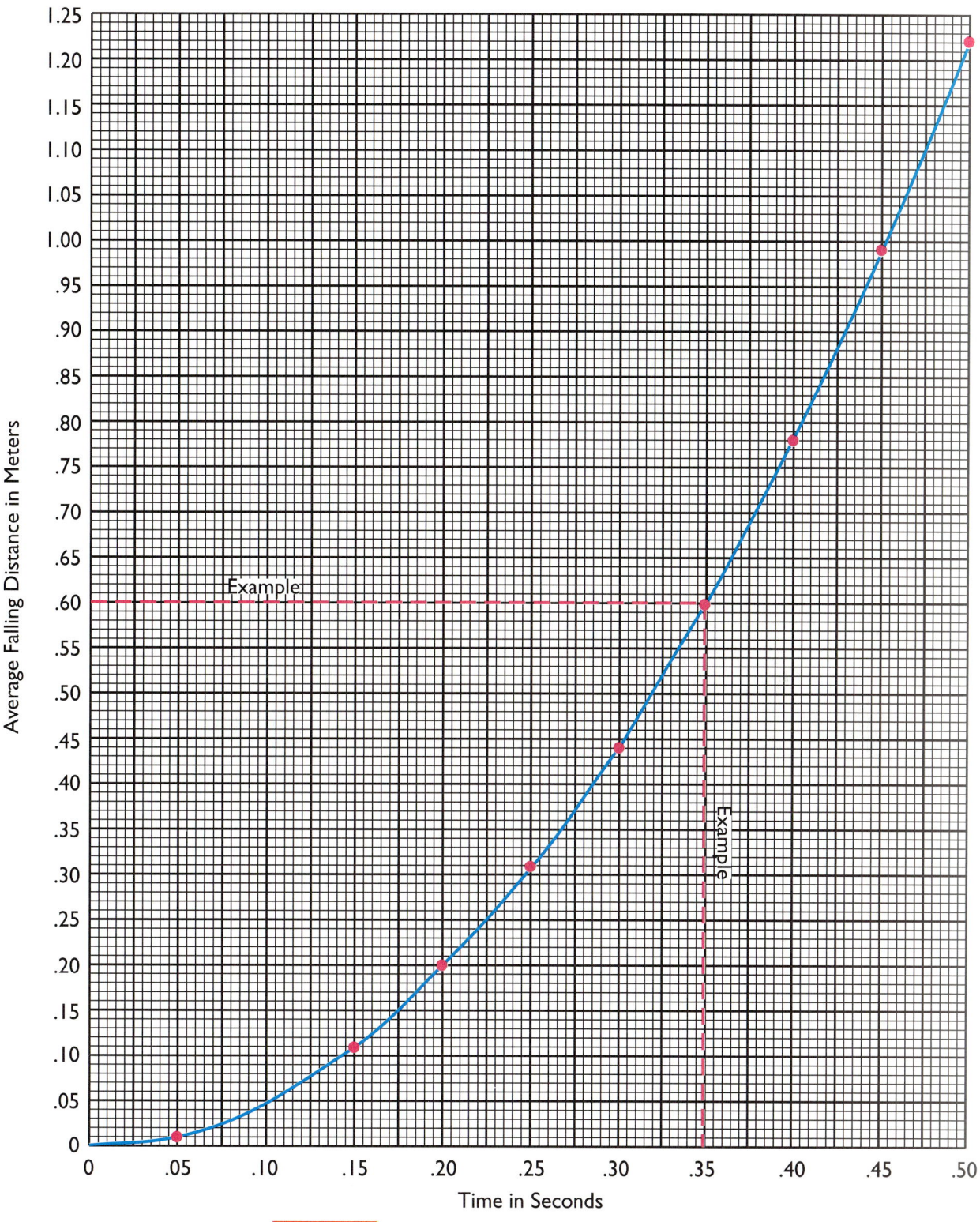

Figure 4.5

This graph shows the relationship between time and average falling distance.

Elaborate — Using Diversity to Set Standards

3. What does the class graph of reaction times resemble?
4. Why does the class graph have the shape it does?
5. What is the range of reaction times for your classmates?
6. Which reaction time is shared by the most people?
7. Based on the class graph, how does your reaction time compare to the reaction times of your classmates?
8. Describe whether or not you think your reaction time is normal.
9. How far would the meter stick fall if your reaction time was 0.45 seconds?
10. Describe whether or not you think a reaction time of 0.45 seconds is normal.
11. Rate yourself on how well you are staying with your group. Choose a rating of A+ for excellent, A for much improved, B for a little better than before, or C for needs improvement. Write your rating in your notebook, using a complete sentence.

INVESTIGATION:
Determining Reaction Distances and Perception Distances

How far you travel while you react, your reaction distance, depends on how fast you are moving and your reaction time. Before you can react to something, you must first perceive it. How much difference is there between reaction and perception distances when you are moving 60 miles per hour compared with when you are moving only 10 miles per hour? By the end of this investigation, you will determine the answers to questions such as these.

Working Environment

You will be working cooperatively in your teams of two. You will need the roles of Team Members, Communicator, and Tracker. Work on the social skill Stay in your groups. Move your desks together side by side.

Materials

For each team of two students:
- 1 ruler

Procedure

1. Read the Background Information following the procedure.

Oh, I see. The times on the horizontal axis stand for perception time or reaction time — whichever we're using! This isn't so bad!

Yeah, this should be fun! Stay in your group, Al....

Hey! Is that ME there in the book?

Elaborate

2. Use the Graph of Reaction and Perception Distances (Figure 4.7) in the Background Information to determine how far you would travel in 1.2 seconds if you were moving 40 miles per hour.

   The distance is 71 feet. This is just practice.

3. Copy the Data Table for Reaction and Perception Distances in your notebook.

   See Figure 4.6.

4. Refer to the class graph of reaction times from the previous investigation. Locate the low, average, and high values.

   Notebook entry: Record these values in the data table.

5. Use the low value to determine the reaction distances for each of the speeds listed in the data table.

   Find the reaction time you're working with on the horizontal axis. Then determine the distance for each speed line on the graph. These distances are the reaction distances.

6. Record the reaction distances in the appropriate column for each of the speeds in your data table.

7. Repeat steps 4 through 6 using the average reaction time.

8. Repeat steps 4 through 6 using the high reaction time.

9. Determine the perception distance for each of these speeds listed in the data table.

   Remember that the average human perception time is 0.75 seconds. On the vertical axis, find 0.75 seconds and use this time to determine distances for the various speed lines. These distances are now the perception distances.

10. Record the perception distances in the appropriate column for each of the speeds in your data table.

Background Information

In this investigation you will use the graph shown in Figure 4.7. This graph is different from the graphs you have used in other investigations because it contains more than one line. Each line represents the data for a different speed limit. For example, the line marked "30 miles per hour" shows the number of feet you travel per second if you are in a car going 30 miles per hour. Graphs such as this are an efficient and convenient way to show large amounts of information.

For practice, use the graph to determine your reaction distance at 30 miles per hour. First, locate your reaction time on the horizontal axis. Next, trace the line directly up from that value until you reach the line labeled "30 miles per hour." Then trace the line to the left from this point until you come to the vertical axis. The value recorded there is your reaction distance at 30 miles per hour.

DATA TABLE FOR REACTION AND PERCEPTION DISTANCES

Speed	Distance with Fastest Reaction Time	Distance with Average Reaction Time	Distance with Slowest Reaction Time	Perception Distance
5 miles per hour				
10 miles per hour				
15 miles per hour				
20 miles per hour				
25 miles per hour				
30 miles per hour				
40 miles per hour				
50 miles per hour				
55 miles per hour				
65 miles per hour				
80 miles per hour				
100 miles per hour				

Figure 4.6

Copy this table into your notebook and fill it in according to steps 4 through 10 in the procedure.

Elaborate

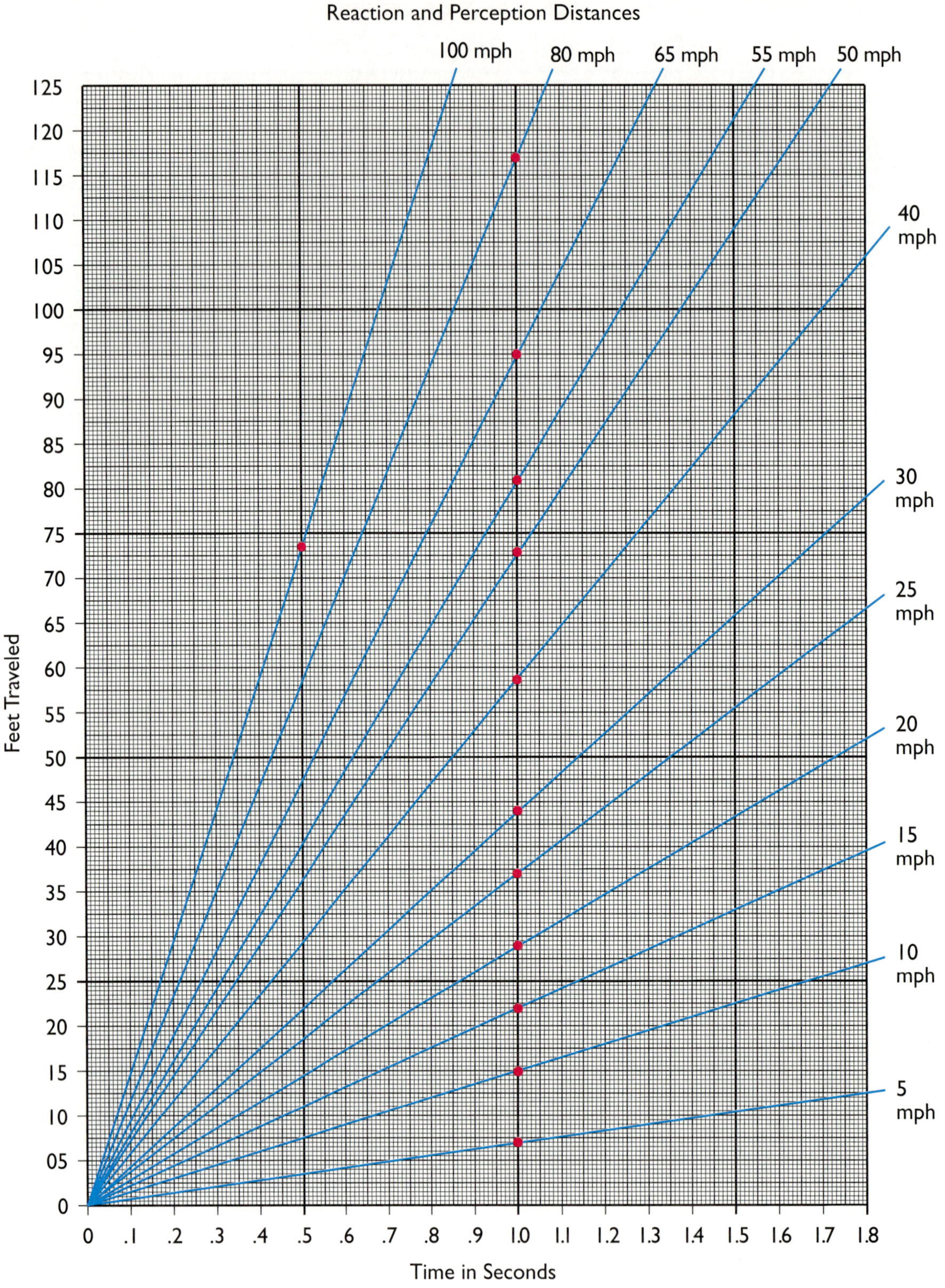

Figure 4.7

This graph shows the distance in feet someone would travel at different speeds and with different reaction and perception times.

Wrap Up

Discuss these questions with your partner and record your answers in your notebook. Be ready to use your answers in the following connections section.

1. Describe how your reaction distance changes as your speed increases.

2. Determine the difference in reaction distance between the *high* and *low* reaction times in your class for the following speeds:
   a. 10 miles per hour
   b. 30 miles per hour
   c. 65 miles per hour

3. Determine the difference in reaction distance between the *high* and the *average* reaction times for the following speeds:
   a. 10 miles per hour
   b. 30 miles per hour
   c. 65 miles per hour

4. How might differences in reaction time influence whether you have enough distance in which to stop to avoid an accident?

5. Tell your teacher whether you feel you have mastered the social skill of staying with your groups.

CONNECTIONS: Total Stopping Distances

Work with your partner on this connections. The reading The Three Phases of Stopping described how you can add together the perception distance, reaction distance, and skidding distance to get the total stopping distance. In the investigation Determining Reaction Distances and Perception Distances, you observed how the distance you will travel changes as your speed increases. In this connections activity you will calculate the total stopping distance for several different speeds.

Copy the Data Table for Total Stopping Distances in your notebook (see Figure 4.8). Transfer the information you need from your Data Table for Reaction and Perception Distances to this data table. Then add the data across each speed limit to determine the total stopping distance for each speed. This means you should choose a reaction distance column from your data table that would result in the safest total stopping distances. Then write answers to the following questions in your notebook. Prepare to share your answers with the rest of the class.

DATA TABLE FOR TOTAL STOPPING DISTANCES

Speed	Reaction Distance (in feet)	Perception Distance (in feet)	Skidding Distance (in feet)	Total Stopping Distance (in feet)
5 miles per hour			1.1	
10 miles per hour			4.7	
15 miles per hour			10.5	
20 miles per hour			19	
25 miles per hour			29	
30 miles per hour			42	
40 miles per hour			75	
50 miles per hour			117	
55 miles per hour			142	
65 miles per hour			198	
80 miles per hour			300	
100 miles per hour			468	

Figure 4.8

Copy this table into your notebook. Fill in the columns according to your calculations on your Data Table for Reaction and Perception Distances. You have to decide which reaction distance column you will use.

1. As speed increases, what happens to total stopping distance?
2. Tell which reaction time you used in your calculations: the class average, the high, or the low. Then tell why you picked the reaction time you did.

INVESTIGATION:
Setting Speed Limits

In this investigation you will use your data from the connections Total Stopping Distances to establish your own set of speed limits. Based on what you know about the relationship between speed and total stopping distance, how would you set and justify speed limits for different locations?

Procedure

1. Decide on speed limits for the following locations:
   a. Divided open highway (interstate)
   b. Rural open highway
   c. School zone
   d. Busy downtown street
   e. Residential neighborhood

Use the information from the Data Table for Total Stopping Distances and consider the following questions for each location:

- *What do you think is an appropriate distance for people to leave between cars in order to be able to stop for an emergency?*

- *The speed limits will be for all people. Should you reconsider your choice of which reaction time to use from the previous connections section?*

- *What are the pros and cons of using the average perception time of 0.75 seconds instead of a slower or faster perception time?*

- *Should you consider road or visibility conditions when setting speed limits? Stopping on wet asphalt instead of dry asphalt can double the skidding distance. Stopping on an icy road can increase the skidding distance by a factor of 11. Traveling downhill increases the stopping distance, and traveling uphill decreases the stopping distance.*

- *How many of the cars on the road are new and in good working condition? How many of the cars on the road have bald tires, wipers that don't work, or brakes that need repair?*

STOP: This is a good time to practice your unit skill. Review the unit skill T-chart if necessary.

Working Environment

Work cooperatively as a Team Member with your partner. One of you also will be the Communicator. When your teacher says "Go," try to be the quickest and quietest team to move into its group. Practice this skill at the end of the investigation as well. Place your chairs so that you are sitting beside each other.

2. Record your team's speed limit decisions.

   Notebook entry: Record these in your notebook beside the speed limits you set in the Wrap Up during the investigation Watching The Final Factor.

3. Be prepared to present and defend your speed limit standards to the rest of the class.

Wrap Up

Discuss the following questions with your partner and write your answers in your notebooks. Prepare to share your answers in a class discussion.

1. Did you decide to set the same speed limit for all areas?
2. Explain the speed limits you set. Tell what factors you considered as you made your decisions.
3. What difficulties did you have in setting speed limits?

4. Listen to your teacher read the current legal speed limit data. Discuss how these data are different from your data and other teams' data and explain why you think this is so.

5. Is your team satisfied with your progress in moving into (and out of) groups quickly and quietly? Explain your answer.

READING:
The Normal Curve and Setting Standards

Setting standards is not an easy thing to do. As you learned in Chapter 3, one type of standard is a rule or limit that society sets. Such rules often are for the safety and well-being of society. But people in a society are very diverse. When you try to set one standard that accounts for human limits and diversity, you are faced with many difficult decisions.

For example, you just completed an investigation in which you set standards—speed limits. You had to decide on one speed that accounted for people's diversity in their ability to perceive and react to emergencies. You might have used 0.75 seconds as the average perception time for humans. You might, however, have adjusted that value to accommodate people who can't perceive sudden events as quickly as others.

You discovered that in your class alone there was a wide diversity in reaction times. Imagine how much wider that diversity would be if you considered the entire population of our country. You had to decide which reaction time you would use in setting your speed limits: the highest, the lowest, or an average.

Stop and Discuss

1. If you picked the average, would your speed limit endanger the lives of all the people whose reaction times are slower than the average?

Not only did you have to consider human limits, but you also considered automobile conditions and weather conditions. You might have assumed that all cars on the road were in a safe driving condition, or you might have assumed that all cars on the road were in very poor condition.

Stop and Discuss

2. Assume that the condition of cars on the road is between average and superior. Would you create potential hazards by setting speed limits too high for the cars that are in the poorest operating condition?

3. You might have set speed limit standards assuming that the roads were dry and visibility was good. Are the roads a safe place to be when the surface is slick and the visibility is poor?
4. What if the people on these slick roads are still driving at the speed limit because they assume that the speed limit was set to account for bad weather conditions?
5. What might happen if you use the averages or high values of human limits when setting standards?
6. What might happen if you use the low values of human limits?

These are all issues that the state and federal governments must keep in mind when setting legal speed limits. There is a way to minimize the problems associated with basing speed limits on just the low, high, or average values of human limits. You could use a normal curve. In using a normal curve, you could examine the way people's limits are distributed. You could decide to use the areas of the curve that account for the most people without any of the extreme values in limits that account for a minority. Figure 4.9 shows different ways you can divide a curve for setting standards.

In fact, that is the way many organizations set standards. They generate a normal curve for a set of data. There is no set method of dividing the curve, however. People who are trying to set standards that they hope will be adequate for an entire population of people must make an important decision. They must decide how to section the data on a normal curve to determine the range of limits for which they want to account.

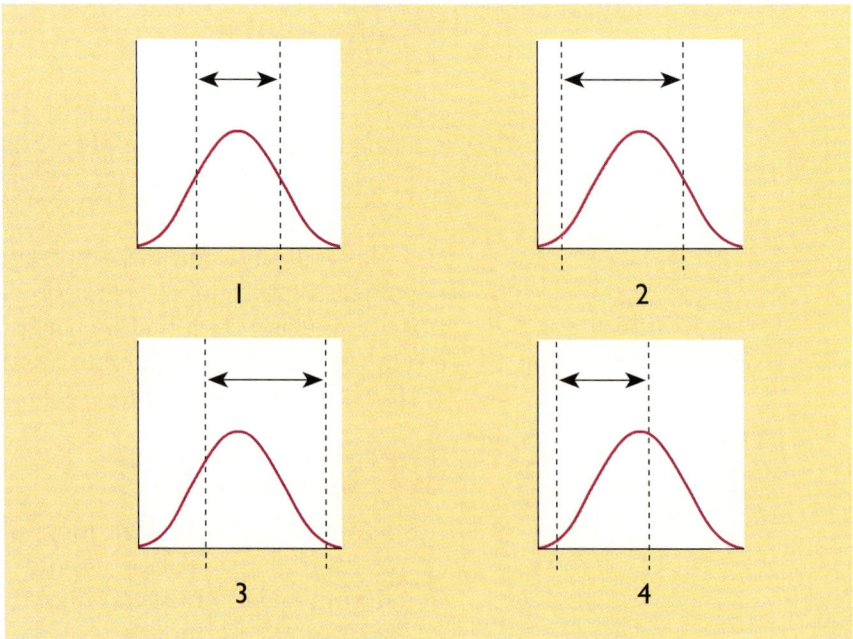

Figure 4.9

There are many ways you can section off a normal curve to account for a majority of people. Can you think of others?

For example, you probably know that a person has to pass a vision test before obtaining a driver's license. If a person cannot see well enough, that person will not receive a license. What does "well enough" mean? Data from a normal curve indicate that some people have poor vision (the low values of the curve). When officials set the standards for eyesight, they may have used this normal curve to section off all the people to the right of the low values and say that they are eligible for a license. Although it is normal for individuals to have poor eyesight, there is a point at which officials determine that vision is too poor to be considered acceptable for driving. Drivers with poor eyesight would be a hazard to themselves and other people on the road.

Although diversity is normal, society often determines that there are ranges of behaviors or limits that are acceptable for certain situations. You could use the normal curve to decide where to section off the low values for reaction time from the rest of the values. For example, you could look at a normal curve of reaction times and decide what part of the curve represents reaction times that are too long. You could assume that individuals with reaction times that are too long should not be driving. You then could base your speed limit on the longest acceptable reaction time. (There is, however, no specific law banning people with very long reaction times from driving.) This is one way you could have used your normal curve data to determine speed limits. Remember that no one can tell you where to section off your curve or what decisions or assumptions you should make. It is up to the person or persons setting the standards to make their own decisions.

Referring to the normal curve is one of the methods people use to set standards. Most decisions that officials make based on data from a normal curve are for the well-being of most people. If you are curious and have some time, you might research how the government makes decisions on speed limit laws and why officials decide what they do. Whether you do that or not, you have experienced how difficult it is to make such decisions. You also have experienced how much people's well-being depends on the decisions of other people.

INVESTIGATION:
Don't Drink and Drive

In the investigation Setting Speed Limits, you set speed limits based on the total stopping distance required at a certain speed. You considered diversity in reaction times, inclement weather conditions, road conditions, and car maintenance conditions, among other factors that could affect speed limit standards. One factor that you probably have heard a great deal about, but that you might not have considered in setting your speed limits, is alcohol consumption. In this investigation you will learn about how alcohol affects people. You then will use this information to set standards for drinking and driving.

Materials

For each team of two students:
- 1 sheet of graph paper
- 3 different colors of pencils
- 1 ruler

Procedure

1. Read the Background Information following this procedure.
2. Use the information in Figure 4.10 to answer the following questions.

   Notebook entry: Write the answers in your notebook.

   a. What do the data suggest about the relationship between people and BAC levels?

   b. Why do you think these three experimenters obtained different results?

   c. Graph the data in Figure 4.10.

   First decide what information you should plot on the horizontal axis and what you should plot on the vertical axis. Be sure to consider how

Working Environment

Work cooperatively in your teams of two. You will need a Tracker and a Communicator. Practice the social skill Stay with your group. Move your desks together or sit beside each other at a table.

Percentage of Test Subjects (People) Who Became Drunk

BAC	Widmark	Jetter	Hine
0.00 – 0.05	0	10	7
0.051 – 0.10	19	8	18
0.101 – 0.15	31	29	24
0.151 – 0.20	33	36	36
0.201 – 0.25	10	7	8
0.251 – 0.30	5	5	4
0.301 – 0.35	2	1	1
0.351 – 0.40	0	0	2
0.401	0	7	0

Source: Council of Scientific Affairs (1986). Alcohol and the Driver, *Journal of the American Medical Association* 255 (4) 522-527.

Figure 4.10

This chart represents the data of three different experimenters (Widmark, Jetter, and Hine) who conducted research on Blood Alcohol Content levels.

to scale each axis. Use a different color pencil for each experimenter's data. Smooth out the bars with a curved line using the appropriate color of pencil.

   d. Decide on a legal standard for driving under the influence (DUI).

   Use the background information on BAC and the graph of the data. Your team should decide at what BAC, and above, it would be illegal for a person to operate a car. You should decide whether you will set just one standard, or whether you will set different standards for males and females, for adults whose stomachs are empty and those whose stomachs are full, and for adults who are small, medium, and large.

3. Prepare to share and defend your decision with the rest of the class.

Be sure that both members of your group can explain how you made the decision and why.

Background Information

There are laws against driving under the influence of alcohol. Officials base these laws on current research that tells them how much alcohol a person can tolerate before becoming a dangerous driver. The people who set the standards and make the laws are concerned about the safety of drivers. They usually base these Driving Under the Influence (DUI) standards on something known as Blood Alcohol Content or BAC. BAC is the ratio of alcohol to total blood volume. This ratio is expressed as a percentage. The range of BAC values is from 0.00 percent, which means there is no alcohol in the blood, to about 0.50 percent, which is usually a fatal amount of alcohol in the blood.

People show a diversity in their tolerances to BAC. Some people might act very drunk with a BAC of 0.05 percent, while others might not act very drunk until they have a BAC of 0.20 percent. One of the things that happens to people who are drunk is that their reaction times increase. People show a wide range of diversity in how much their reaction times increase at different BACs.

People also show a wide range of diversity in BAC after drinking the same amount of alcohol. A smaller adult who drinks as much as a larger adult will have a higher BAC, because the smaller adult has less blood. One study has shown that men and women differ in how they digest alcohol. According to that study, a woman who drinks the same amount of alcohol as a man will have a higher BAC than the man, even after allowing for differences in size.

Also whether a person has a full or empty stomach affects his or her ability to absorb alcohol into the blood. Because food in the stomach absorbs some of the alcohol before it gets into the blood, the person with a full stomach will have a lower BAC than the person with an empty stomach.

Figure 4.11

Police need a convenient way to test for alcohol consumption on the side of the road. They use an instrument called a Breathalyzer™, which measures the amount of alcohol in one's breath.

Wrap Up

Discuss the following questions with your partner. When you agree on an answer, write the answer in your notebook and prepare to discuss it with the rest of the class.

1. How were the standards other groups set similar to yours?
2. How were the standards other groups set different from yours?
3. Calculate one DUI standard for the entire class. (You can average the teams' standards, or you can obtain a standard from a graph of the class's standards.)
4. You based your DUI standards on data that told you at what BAC levels people became drunk. What is your operational definition of drunk? (Relate your operational definition first to driving, then to walking, then to riding a bike.)
5. How do you suppose Widmark, Jetter, and Hine would operationally define "drunk"?
6. How might the operational definitions of Widmark, Jetter, and Hine differ?
7. In your notebook write a short paragraph that describes how you have improved in your ability to stay with your group.

CONNECTIONS:
Drinking and Driving—Your Decision

After each team has presented its DUI standard, consider and answer the following as a class.

Imagine that your class was the state legislature. Would your DUI laws be stricter or more lenient than the current DUI laws for your state? Explain how they would be different and why. Do you feel safe with the current DUI laws for your state?

CHAPTER 5

Evaluating Your Understanding of Limits and Diversity

Congratulations! You almost have completed the first unit in this book. That is quite an accomplishment! You have explored and learned many things about limits and diversity since the investigation Star Tracers. The next reading should help you put it all in perspective.

READING:
What Can You Do with Unit 1?

You began Chapter 1 by conducting investigations in which you measured your classmates' successes at accomplishing certain tasks. You quickly found out that people have limits in accomplishing certain tasks and that people show diversity in their limits. Your investigations yielded data, which you learned to organize into data tables and which you also learned to graph. The shapes of the graphs were similar to one another. After you accomplished all of this, you also learned how to collect data that could be compared to other teams' data by identifying common operational definitions and controlling variables.

In Chapter 2 you used popcorn to learn that being *different* is what is normal. Normal doesn't mean being able to do things the same as everyone else. You learned this by collecting and studying data that generated a normal curve and by studying curves of humans' and animals' peripheral vision. You were able to practice your skills of creating operational definitions and controlling variables when you explored the concept of afterimage and discovered more diversity in your class.

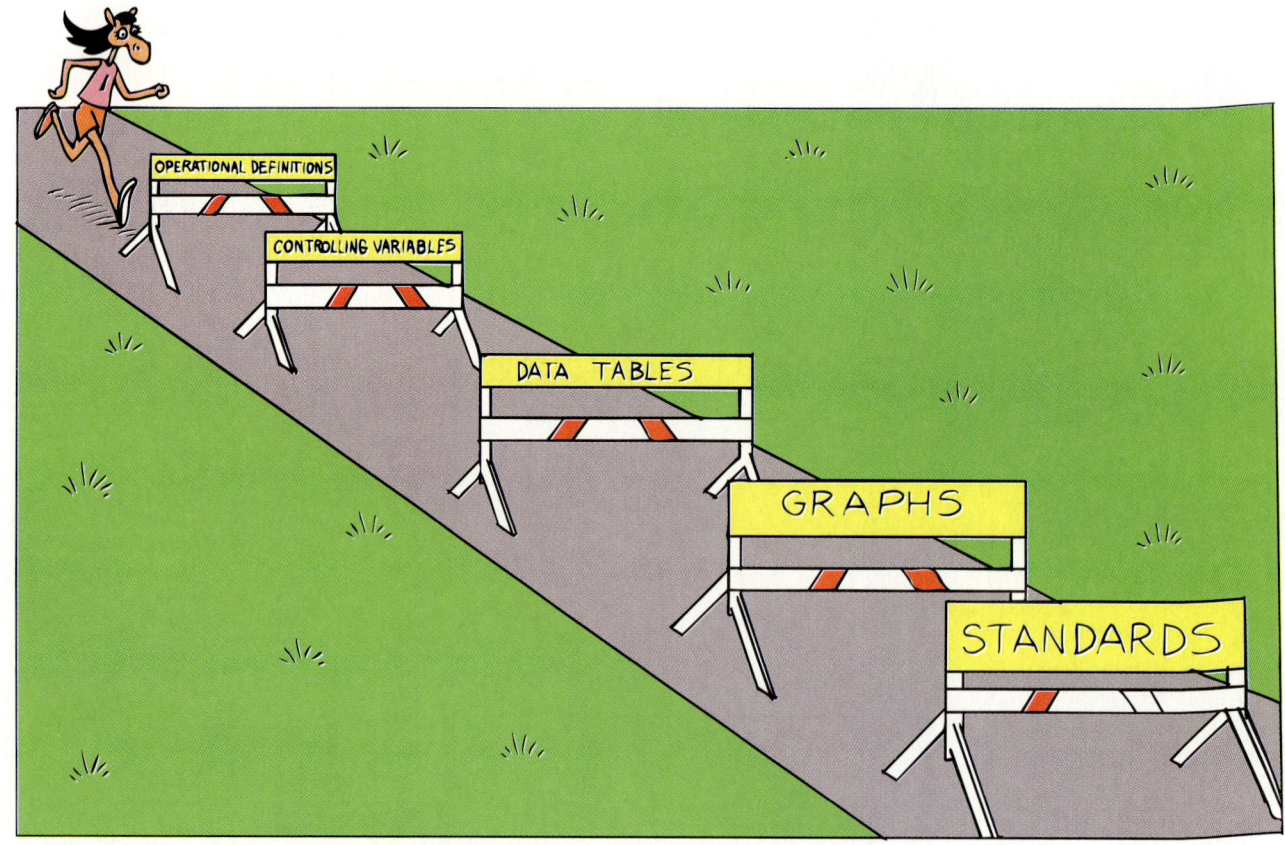

In Chapter 3 you had a chance to use TV to identify more limits in humans and to find that humans are diverse in their ability to perceive continuous motion, flicker, and lines at a distance. You also learned a lot about television. You set standards for TV pictures that accounted for human limits.

In Chapter 4 you explored the three phases of stopping and calculated the total stopping distance for people traveling at certain speeds. You used the diversity in people's ability to react to sudden events and considered those limits and other factors that affect stopping distance. This helped you set standards for speed limits in different areas of the city and country. You learned how using a normal curve could help you make important decisions about setting standards, and you practiced using a normal curve to help you set standards for drinking and driving.

If Chapter 5 were an Olympic event, it might be called "Unit 1 Marathon," because you are going to take everything you studied in the first four chapters and use it in one investigation in this chapter. You will create operational definitions, control variables, and write a procedure that will help you conduct an experiment and collect data from 25 people. After organizing your data, you will graph it and then use your graph to set a standard. When you are finished with that investigation, you can compare it to Star Tracers and get a good feeling for what you have accomplished so far.

INVESTIGATION:
How Much Noise Is Too Much Noise?

You are competing for the prestigious honor of being named to the Noise Patrol Squad at your school. Your school has just lost an entire team from the Noise Patrol Squad because the Thompson triplets have moved away. Your team is competing against all the other finalists that you see in your room. (Don't look now; they might think that you are checking out the competition.)

The Noise Patrol Squad has given you the following definition of "too noisy": It is too noisy when the noise level in the room interferes with a person's comprehension when reading. Your team must design an experiment that will use this definition of "too noisy" to find out the diversity of tolerance to noise in a population of 25 people. The procedure section outlines the requirements of this contest. Read through the requirements first so that you have a better understanding of what you will be asked to do. If you are unsure of what the Noise Patrol Squad wants you to do, have your Communicator ask for help from the judge for the Noise Patrol Squad.

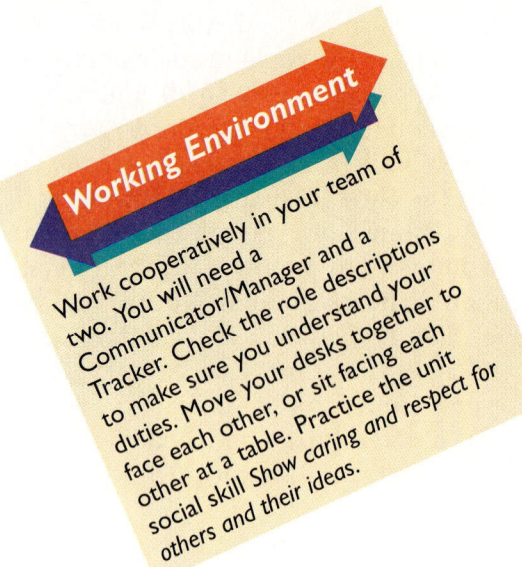

Working Environment

Work cooperatively in your team of two. You will need a Communicator/Manager and a Tracker. Check the role descriptions to make sure you understand your duties. Move your desks together to face each other, or sit facing each other at a table. Practice the unit social skill Show caring and respect for others and their ideas.

Materials

For each team of two students:
You will determine what materials you need. (The materials you need might change as you gather results.)

Procedure

1. Have a brainstorming session to create operational definitions. One operational definition should say how you will measure a person's comprehension while reading. The other operational definition should say how you will measure the loudness level that first interferes with a person's reading comprehension.

   Notebook entry: The Communicator should record all of the ideas in his or her notebook. Be careful not to rule out any ideas yet.

2. As a team, decide which operational definition is the one your team will use while conducting your investigation.

   Use the list you generated in step 1.

3. Create a list of variables that you will need to control while conducting your investigation.

   Notebook entry: Record these in your notebook.

4. Identify a list of materials and a procedure that your team will use to conduct your investigation.

   Notebook entry: Record these in your notebook.

5. Design a data table to help as you collect your data.

6. Conduct your investigation.

   Make sure to follow your procedure, use your operational definitions, and control your variables. Collect data from any 25 people you choose. If you find that your operational definitions are not working, you may develop new ones.

7. Construct a graph of your results.

8. Use the pattern that your graph makes to determine the range of tolerance to noise among your 25 test subjects.

9. Set a standard noise level for the classrooms in your school by completing this sentence: "It is too noisy to read when the level of noise in the room exceeds _____."

   Base this standard on the diversity among your test subjects. Remember, because your team is deciding on the standard level of noise, only your team can determine what, if any, cut-off points on your graph there might be.

Wrap Up

Prepare a presentation for the Noise Patrol Squad. Begin your presentation with the sentence you completed above ("It is too noisy to read when the level of noise in the room exceeds _____."). Also include the following in your presentation:

- your operational definitions and how they worked for your team,
- the variables you had to control and how you controlled them,
- the data you collected,
- the graph that your data generated,
- how you used the graph or graphs to determine a tolerable noise level,
- what a tolerable noise level is for studying in classrooms in your school,
- specific examples of how each member of your team practiced the social skill, and
- your team's overall social skill rating (use a scale of 1 to 10, where 10 is perfect).

Keep your objective in mind: Win the coveted slots on your school's Noise Patrol Squad! Because you are part of a team, you may find it easier to make your presentation if each of you prepares two or three of the sections listed above.

SIDELIGHT

Shhh...Hush...QUIET!

Which of these noisy things do you consider too noisy?

UNIT 2

How Does Technology Account for My Limits?

In Unit 1, you explored the limits that you and your classmates experienced in accomplishing certain tasks. You found that all humans have different limits. At times we would like to change some of our limits. For example, some people wear glasses or contact lenses. These devices allow people to change the limit of how well they can see close up or at a distance. What advantages are there to changing the limits of your sight? One advantage is seeing better so that you can function more easily in the world.

In this unit you will have a chance to explore technological design. This is an aspect of science that is devoted to changing and improving human limits. First you will learn some basic concepts about design. Then you will explore some familiar objects that help humans overcome transportation problems as well as other objects that help humans overcome boredom. You also will find out how people design these and many other objects by accounting for human factors. You then will study the technological design process, which is all about designing objects that help people overcome limits and solve problems.

COOPERATIVE LEARNING OVERVIEW

In Unit 1 you used cooperative learning to accomplish many tasks. You might have become fairly comfortable working with your partner. You also might have found that practicing the skills became easier as the unit progressed. But now that you are starting a new unit, it is time to form new teams. This time you will work cooperatively in one team of three for the entire unit. Some of you might feel like Ros; she's ready to move on and try a new team. Others of you might identify more with how Isaac feels; he's a bit uneasy with the idea of new teams, or at least is hoping that he can choose his own teammates. That's okay. However you feel, you should try to remember the skills you practiced in Unit 1, especially the unit skill. Apply these skills to your new team situation. Continuing to practice these skills should help you make an easier transition into a new team.

As you keep the Unit 1 skills in mind, practice the new skills in this unit. Also, practice the Unit 2 skill of being open to others' ideas throughout the unit, whether the Working Environment mentions it or not. Apply this new unit skill to class discussions, presentations, and other situations in which you must work outside your team.

Take time to review the role descriptions again. Discuss the times when you felt that the duties were not clear. For example, when can someone besides the Manager obtain materials? When does someone other than the Tracker do the timing? What is the best way for the Communicator to seek help from the teacher?

Finally spend a few minutes discussing with your classmates the Unit 2 skill of being open to others' ideas. Do this together by constructing a class T-chart for the skill. Make a copy of the T-chart for yourself in your notebook.

CHAPTER **6**

Consumer Concerns

If you watch **TV**, then you are used to seeing a lot of commercials. Manufacturers spend about $150,000–$200,000 per minute on television commercials to convince you that their product is the best. But are all products as good as their manufacturers claim? How can each brand of paper towel be the best? How can each type of breakfast cereal be the tastiest or the healthiest?

Usually advertisers create catchy slogans or memorable commercials to convince consumers (the people who buy products) to buy *their* products. Another technique that advertisers use is to provide free samples or coupons so that people will try a product. Have you ever chosen a breakfast cereal because the prize inside seemed really neat?

These advertising techniques might get consumers to *buy* the product, but this doesn't necessarily mean that the product the consumers buy *is* the best. Nor does it mean that the consumer will *continue* to buy the product. In this chapter you will investigate paper towels and breakfast cereals and learn to become better consumers. You also will investigate how to determine which products are the best.

INVESTIGATION:
Tall, Dark, Handsome, Strong, and Absorbent

Who makes the best paper towel? How can you find out? This investigation will help you answer these questions. You will test only a few brands of paper towels. Once you figure out how to test a few, you can be a smart consumer and test other brands on your own.

Materials

For the entire class:
- 1 pan balance
- 1 to 1.5 kg of washers, pennies, or other uniform objects

For each team of three students:
- 1 tray
- 1 metric ruler
- 1 beaker or measuring cup
- 1 sample of each brand of paper towels your teacher provides
- a water source
- any other supplies you think you might need from those your teacher provides

Working Environment

Work cooperatively in your new teams of three. The roles you will need are Manager, Communicator, and Tracker. (Remember that each of you should always fulfill the duties of Team Members.) Practice the social skill Use your teammates' names. Move your desks together or sit in a triangular configuration at a table. Make sure you can see each other's faces.

Procedure: Part A—The Social Skill

1. Write your teammates' first and last names in your notebook.
2. With your new teammates, share your Unit 1 ideas about why using each others' names is an important social skill.

   You will find these in your notebooks in the first investigation, Star Tracers. Make sure each Team Member reviews everybody's reasons.

Procedure: Part B—The Activity

1. Obtain the materials.
2. Label each sample of paper towel with its brand name.

   You can do this using initials or symbols in one corner of each paper towel.

3. As a team conduct a brainstorming session to create a list of properties that a "good" paper towel has. For example, is a good paper towel strong, absorbent, or soft?

   Notebook entry: Record this list in your science notebook.

4. Rank the properties on your list from most important to least important.

   Notebook entry: Record this ranking in your notebook.

5. Pick one property from your list that you want to test first in your samples.

   You can test as many properties as you want, but you need to begin with just one property.

6. As a team create an operational definition that states how you will measure each property that you chose.

   Notebook entry: Record your operational definitions. Be creative and try to come up with operational definitions that will not cause you to waste too many paper towels.

7. For each operational definition, list the variables you will need to control to make your tests fair.

   Notebook entry: Record the variables. You might not think of all the variables until you are performing your experiment. That's okay. You can record others as you work.

   STOP: Are you being open to the ideas of your teammates? Are you using each others' names?

8. Construct a data table for your tests.

   In this data table, you will record all of your tests and their results. The data table also should include a column for ranking the paper towels from best overall to worst overall.

9. Conduct your investigation.

   In your data table, record any information you need along with the results you obtain.

10. Rank the paper towels from best to worst.

    Base your ranking on the results of your team's tests.

    Notebook entry: Record your paper towel ranking.

11. Record your rankings on the class data table.

    This is the Communicator's job. Your teacher will provide this data table on the chalkboard or overhead transparency.

Wrap Up

Discuss the following with your team. Record your responses in your notebook. Be sure each member of your team can explain the team's operational definitions and the properties that your team decided were important to test.

1. Compare your list of rankings with other teams. Did any teams have identical rankings?
2. Describe a reason why there might be a difference in the rankings among teams.
3. Have your Communicator compare operational definitions with other teams. Did any teams create the same operational definitions?
4. If some teams used the same operational definitions, did these teams rank the towels identically? Why or why not?
5. What are your teammates' names?
6. How much improvement in using each others' names does your team need: none, 50 percent, more than 75 percent?
7. Each Team Member should recall and record one time he or she was open to another's idea. Then discuss how being open to others' ideas is different from showing caring and respect for others' ideas.

READING:
Paper Towel Consumers

The following is an article reprinted from Consumer Reports, a magazine that provides product information and advice to consumers about purchasing products and services. The magazine staff ranks products just as you did in the previous investigation. We have inserted questions at various places in the article to help you think about specific things. Read aloud each section in your team of three by having the Manager read the section before questions 1 and 2. The Communicator then should read the section after that, stopping before question 3. Then the Tracker should read the section before questions 4 and 5. Divide the remaining sections into parts and continue to take turns reading in the same order. When you get to Stop and Discuss, discuss the questions. Write answers to the questions in your notebook.

Some costly brands outperformed the ordinary towels, but that doesn't make them better values.

Paper towels are a pretty humdrum product. One supermarket executive told us "they're the closest thing to cordwood my store sells." This isn't because of the shape or the way the towels are stacked, but because the retailer treats one roll of paper pretty much the same as another.

The manufacturers, however, go to some lengths to persuade you that paper towels are anything but cordwood. Different colors and decorator designs compete for your attention on the shelves.

Some manufacturers play with the packaging, wrapping two or three rolls together to make you think the bigger pack is a better value than a single roll. Georgia-Pacific's *Mr. Big*, for one, is sold only in a three-roll pack. You can buy most other brands one roll at a time.

Stop and Discuss

1. Explain the importance of color and decoration to manufacturers and consumers.
2. As a consumer how do you feel about manufacturers trying to make three-roll packages appear to be a better deal? If you were a manufacturer, how would you feel about doing that?

Because paper towels are costly to haul long distances, most big paper towel companies make their towels at different mills. Some brands are the same nationwide, but there are also many regional and store brands. In a few cases, a nationally known brand name will vary from region to region. For example, the *Viva* we bought in the Midwest and West is a one-ply towel; the East Coast *Viva* is a two-ply towel. *Hi-Dri* has 100 sheets to the roll on the West Coast, 200 sheets in the East.

Stop and Discuss

3. Can you think of any reasons for having different packaging in different parts of the country? How do manufacturers find out about these reasons?

The manufacturers try to take a bigger share of the market by selling many brands, pitching them to different segments of the market. Scott Paper Co., for example, sells *Job Squad* and *Viva* at a premium price, aiming those brands at consumers who believe that a high price connotes high quality. Scott Paper also sells *ScotTowels*, a moderately priced brand aimed at the consumers who treat towels like cordwood.

Some manufacturers split the market even finer. Proctor & Gamble sells two kinds of *Bounty,* a "regular" towel and one meant for use in a microwave oven. The company wants you to believe that you actually need two kinds of *Bounty* in the kitchen. Scott Paper is trying to segment the market on size; its *ScotTowels Junior,* an 8-$\frac{1}{4}$-inch wide roll, is pitched as a towel that's "just the right size for saving money." (It's not.)

Stop and Discuss

4. Describe a time when you or someone you know bought something expensive because you or they thought it was better than an inexpensive brand.

5. What problem do you see with the 8-$\frac{1}{4}$-inch size of *ScotTowels Junior* if all other paper towel rolls are 11 inches wide?

The supermarket executive we spoke with termed the premium-priced towels "overspecified," meaning that they are thicker and heavier than they have to be. The overspecified towel gives the advertiser something to brag about and helps justify the generally higher price. That in turn pays for both the manufacturing costs and the heavy advertising and promotion expenses. Proctor & Gamble, the leading advertiser, has spent millions to have Rosie the waitress tout the strengths of premium-priced *Bounty* as "the quicker picker-upper."

The well-rounded towel

Our tests of paper towels were something of a departure from Rosie's demonstrations in the *Bounty* ads. But, like Rosie, we looked for towels that had both good wet strength and good absorbency. Combining those qualities is something of a technical achievement. Most of the papermaking processes that create strength tend to undercut absorbency and vice versa.

To test for absorbency, we weighed dry towels, dipped them in water, skimmed them across the lip of the pan, then weighed them again to see how much water they picked up. In general, the premium-priced towels, both one- and two-ply, were more absorbent than the lower-priced major brands or store brands. The better towels have more of what the trade calls "puff."

Absorbency alone doesn't help much if it takes the towel a long time to get wet. If you're trying to mop up a large spill, or if you spill something on a carpet, you'll want a towel that absorbs liquid fast. To see how quickly the towels could absorb, we ran speed trials using both water and cooking oil.

With the towels clamped in an embroidery hoop, we dripped water and oil on them in separate tests, noting how long it took for the liquid to soak in. Several towels—store brands as well as premium brands—soaked up water almost instantaneously; the slowest took about 20 seconds to absorb the water drops.

All the towels absorbed the oil more slowly. The fastest took two to three seconds to drink a drop of oil and the slowest took a leisurely 2-$\frac{1}{2}$ minutes.

We measured wet strength two ways. First, to find out how much weight a wet towel could bear, we mounted each towel in an embroidery hoop, wetted it, then poured on a steady stream of lead shot. When the towel burst, we stopped pouring and weighed the shot. The weakest towels held only about half a pound of lead; the strongest, more than three pounds. Second, to see how the towels would hold up to scrubbing, we mounted them on a laboratory machine that rubbed them over a textured plastic surface. Some towels disintegrated after about a dozen strokes, while others lasted for a few hundred.

Job Squad proved to be the strongest in these tests. But its high strength—and high price—verge on overkill. The one-ply *Viva* was amply strong, overall, and costs only about half as much as *Job Squad*.

Stop and Discuss

6. What properties of paper towels did the *Consumer Reports* investigators test?
7. Why do you suppose the investigators thought these qualities were important to test?
8. From the previous section titled "The well-rounded towel," list the operational definitions the investigators used.
9. List all the variables you can think of that the experimenters had to control for each test they performed.

Linting and running

Some paper towels shed lint when they're used to wipe a hard surface, a shortcoming that's particularly noticeable if you use paper towels to wash windows or mirrors. The highest-rated towels were about average here.

Towels printed in vivid colors posed another problem: Some of the wet towel's color rubbed off onto white cotton or white-painted panels. Since towel colors and designs change frequently, we can't say which brands are the most likely to run. But you can always play it safe and use white paper towels.

A towel that doesn't separate quite right at the perforations can leave you with an avalanche of paper or a mere shred. We tore hundreds of towels off their rolls to see which came off neatly. Nearly all did, provided the towel holder was properly tensioned. Only the *Delta* gave us a ragged portion of towel from time to time.

Stop and Discuss

10. Explain why the following properties might matter to the consumer: linting, colors rubbing off, and towels not tearing easily at the perforations.

Recommendations

The strongest, most absorbent towels were the premium-priced brands such as *Job Squad*, *Viva*, and *Bounty*. Judged strictly on performance, *Job Squad* and the one-ply version of *Viva* earned a check rating. But that doesn't make those towels the best value. For simple little spills or other small mop-ups, you might want to keep a roll of cheap towels. Look to the Ratings column for unit cost per 100 towels to find the true bargains. For example, store brands such as *Pathmark* (38 cents per 100). Safeway's *Marigold* (51 cents), Kroger's *Cost Cutter* (39 cents), or A&P (59 cents) are low in the Ratings yet adequate for undemanding jobs. By contrast, *Job Squad* costs $1.84 per 100, one-ply *Viva* costs 92 cents per 100, and *Bounty* costs $1.09.

For bigger jobs, such as washing windows or cleaning the cook top, you can always keep a spare roll of towels under the sink or in the broom closet. In this case, though, you should look for a brand that did well in our test for absorbency and wet strength and that's moderately priced by the square foot rather than by the towel. Check the column for unit cost per 100 square feet to find the better values. *Brawny* ($1.05 per 100 square feet) and *ScotTowels* (84 cents) are among the cheaper towels that are also good enough to handle most heavy-duty cleaning chores.

A special towel for a special need?

Procter & Gamble's *Bounty Microwave* is one of those products that solves a problem you didn't know you had. Ever since microwave ovens caught on, people have used paper towels to cook in them. Paper towels keep foods such as bacon or sausage from making a mess of the oven interior; the towels also help keep bread and rolls from drying out or getting soggy when they're warmed.

Bounty Microwave, though it claims to be a towel for all tasks, is cagily named to make you think it's somehow special and therefore better than other brands for microwave cooking. But there's only one meaningful difference we could find between the two types of *Bounty:* We paid a bit more for the microwave version.

Procter & Gamble claims that *Bounty Microwave* contains no artificial colors. In other words, it's white. In our opinion, any white paper towel should work in a microwave oven.

Explore

A cents-off coupon or a special store sale may make an otherwise expensive brand a good buy. But don't think that towels in two- or three-roll packs give you a price break: most of the multiple-roll packs we looked at were no cheaper per 100 towels or per 100 square feet than single rolls of the same brand.

Stop and Discuss

11. Why would you look for a brand of paper towels priced by the square foot instead of by the towel?

Use the Ratings chart to answer the following questions.

12. Which brand(s) of paper towels is (are) the most expensive per roll?

13. Which brand(s) of paper towels is (are) the most expensive per 100 towels?

14. Which brand(s) of paper towels is (are) the most expensive per 100 sq feet?

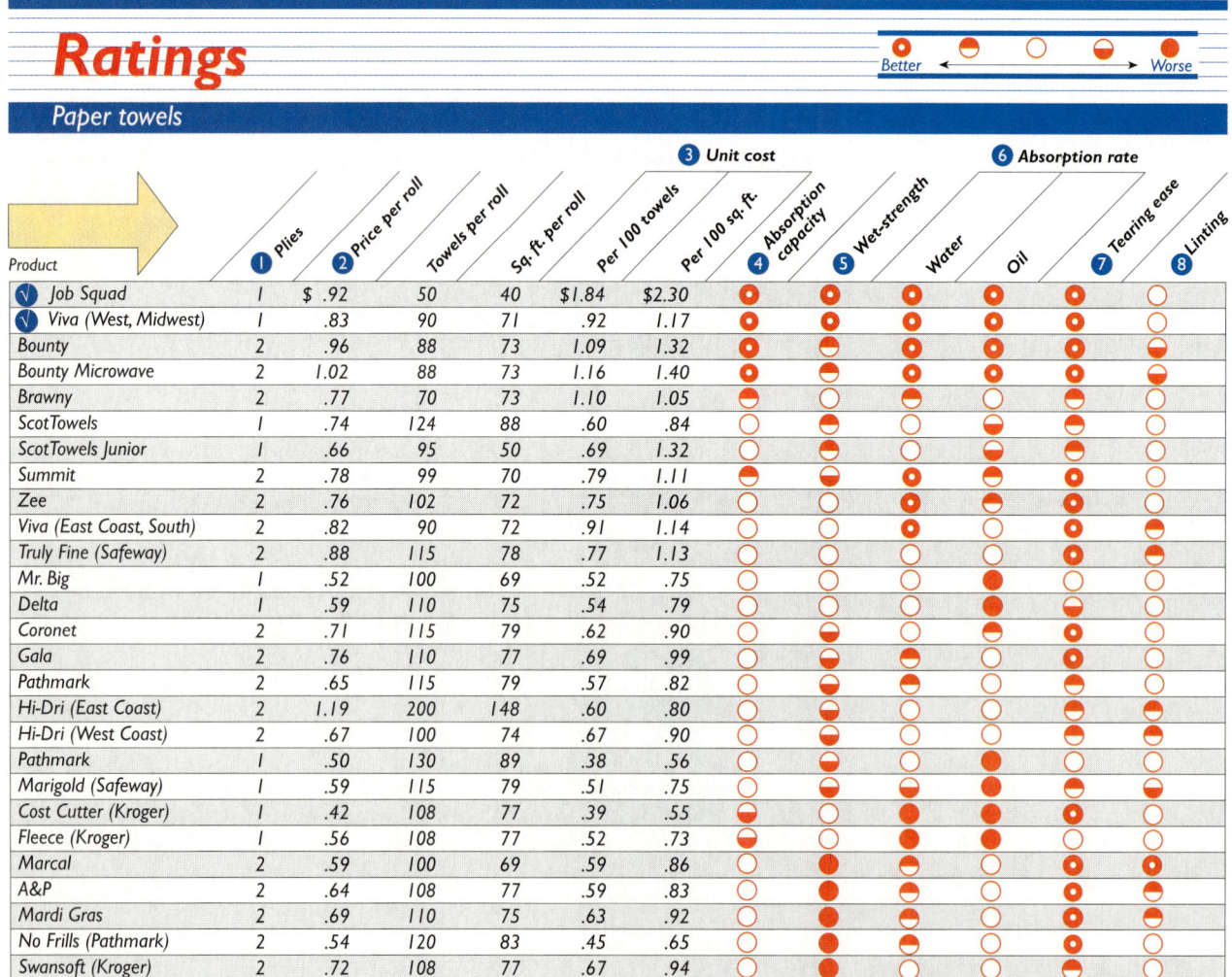

15. How are the *Hi-Dri* paper towels sold on the East Coast different from the *Hi-Dri* paper towels sold on the West Coast?
16. Is the *Cost Cutter* brand of paper towels a better buy than the *Pathmark* brand of paper towels?
17. What operational definition did the investigators use to measure linting (category 8)?
18. Which brand performed better in the wet-strength test, *Viva* or *Gala?*
19. Which brand of paper towels did *Consumer Reports* rank as the best towel?

Copyright 1987 by Consumers Union of United States, Inc., Yonkers, NY 10703. Adapted by permission from *Consumer Reports,* September 1987.

Guide to the Ratings

Listed in order of estimated quality. Differences between closely ranked models were slight.

1. **Plies.** Useful more as an identifying mark than as a sign of quality.
2. **Price per roll.** Averages of the prices we paid for single rolls bought in different areas. The price of *Mr. Big,* available only in a three-roll pack, was adjusted accordingly.
3. **Unit cost.** Shown two ways, **per 100 towels** and **per 100 square feet.** If you use paper towels primarily for small jobs, which require one or two towels at a time, the cost per towel is the more significant factor. If you want paper towels primarily for window-washing and big jobs, the cost per square foot is more meaningful.
4. **Absorption capacity.** This was a key factor in determining the Ratings order. In our tests, the best towels absorbed about four times as much water as the worst.
5. **Wet-strength.** Another key factor in determining the Ratings order. It tells you which towels should hold up in wet cleaning and scrubbing.
6. **Absorption rate.** No matter how absorbent a towel is, that quality does little good if the towel takes too long to soak up a spill, or the grease from fried chicken. We dripped water and cooking oil on the towels and clocked the rate of absorption. The thirstiest soaked up water almost instantaneously and absorbed the oil in two to three seconds. Some of the worst needed a couple of minutes to absorb the oil.
7. **Tearing ease.** Towels that don't separate easily at the perforations are annoying. We found few problems, provided the towel holder was properly tensioned.
8. **Linting.** The lower the score, the more tiny flecks a towel left behind. Linting is especially bothersome if you use paper towels to clean windows or mirrors.

Explore

CONNECTIONS:
Comparing Ratings

Use your data from Tall, Dark, Handsome, Strong, and Absorbent and the *Consumer Reports* article to complete the following. Record your answers in your notebook.

1. Describe whether or not all, most, or none of your operational definitions match the operational definitions that the *Consumer Reports* investigators used.
2. Describe how similar your rankings of the best to worst brands are to the final rankings in the *Consumer Reports* article.

READING:
Why Products Fit

Recall from Unit 1 that you designed and advertised a television set. Before you actually designed the TV, you explored some of the limits of human vision that relate to the design of a TV screen. Later you learned that these limits are called *human factors* and that people design products with human factors in mind. In the process of designing a TV screen, you set *standards* based on human factors. When manufacturers don't account for human factors, their products might not "fit" people very well. That's not good for consumers. If it isn't good for consumers, it isn't good for manufacturers either, because then people won't buy the manufacturers' products.

But what does all of this have to do with paper towels? You didn't design paper towels and the staff of *Consumer Reports* didn't design paper towels. But you did test towels and you chose what properties you would test. As you read the following scene from One Morning at Work, imagine that you are a paper towel designer.

One Morning at Work

Scene: *An office in a high-rise building in downtown New York. All around the office are drafting boards, drawing supplies, paper, and samples of paper towels currently on the market. Four young designers are beginning the task of designing a new brand of paper towels. The designers names are Isaac, Ros, Al, and Marie.*

Ros: If we want the product to sell, I suppose we want our towel to be the best at everything!

Isaac: It can't be the best at everything. We need to narrow it down.

Al: Yeah, if we narrow it down to being the best at some things, we can make a better towel.

Marie: Okay, then let's set some goals for our towel. What do we want our towel to be?

Isaac: The strongest!

AL: The most attractive on the shelf!

ROS: The most absorbent!

MARIE: The most convenient!

AL: Write all that down, someone!

MARIE: I will. (*She takes time to record all the ideas. She titles the list "Goals."*) Okay, any others?

ISAAC: That seems like the basics.

MARIE: Are there any goals on here we don't think are possible?

ROS: They all look good, but we need to know more about each one because that will make a difference in how we reach our goals.

AL: Like what?

ROS: Well, for starters, when people want a strong paper towel, what does that mean? What are they going to use a paper towel for that it needs to be so strong? I mean, who needs a strong paper towel just to dry their hands?

AL: People don't use towels just for drying hands, Ros.

ISAAC: Most people use towels for cleaning.

MARIE: So? I don't need a paper towel to be strong just to wipe smudges off a mirror.

Explain Consumer Concerns

AL: No, people use them for more than that. These days they expect paper towels to be like sponges. They use scouring powder and a paper towel to clean their kitchen sinks.

ISAAC: Great, we'll make a towel so strong it doesn't get holes or shred when you scrub with scouring powder.

ROS: Okay, then what about attractive? What do people find attractive?

MARIE: That probably depends on how they've decorated their kitchens—you know, they want their towels to match their kitchens. Like, for instance you, Al. What would match your kitchen?

AL: A nice black, purple, and yellow plaid with an olive checkerboard border.

ISAAC: You asked the wrong guy, Marie. Somehow, I don't think Al has typical tastes.

ROS: Then we need to decide what typical tastes are.

MARIE: I say we stick with colors and patterns that go with things in most kitchens—like blue, green, gold, and brown.

ISAAC: And let's stick with flower patterns.

AL: Good idea. Are you getting this down, Marie?

MARIE: Is your hand broken, Al?

AL: No, but I wouldn't know what to write!

MARIE: Come over here. Just make a column there beside my "Goals" column.

ISAAC: Call it, "Things that limit goals."

AL: (*He trades places with Marie and begins writing. Beside "Strongest" he writes "People use for cleaning with scouring powder." Beside "Most attractive" he writes, "Must match typical kitchens."*) What do I put by "Most absorbent?"

Figure 6.1

So far, this is what Marie and Al have recorded.

Goals	Things that limit goals
Strongest	people use for cleaning with scouring powder
Most attractive	must match typical kitchens
Most absorbent	
Most convenient	

Ros: At my house most of the spills happen at the dinner table when someone spills a drink, and then everyone bolts for the paper towels.

Al: Yeah, but you end up using about ten paper towels.

Marie: Then write down, "Average spill is about one full glass."

Isaac: A full glass is about eight ounces. Put "8 ounces."

Ros: What about "Most convenient"?

Marie: Well, we once bought a roll of towels that didn't fit our paper towel holder, so we just left the towels lying around the kitchen. Then we could never find them when we needed them. Finally I knocked them into the sink and ruined half the roll.

Isaac: Another thing that burns me is when you're racing to wipe up a spill and you yank at a towel and run and the whole roll comes with you!

Ros: Okay, then put "Size of a typical holder" and also put "Strength of a typical tug."

Marie: You know, you guys, I'm thinking we might get carried away and need too many expensive materials for these great towels we're designing. We better be careful about that. We can design all we want, but it won't do us any good if the manufacturing department says they can't make them because they don't have enough money!

Isaac: Well, that sort of limits our goals, too. We're limited in the materials we can afford to buy. Al, write "Materials" next on the list in the "Things that limit goals" column.

Ros: So people have a big effect on the goals, but other things do, too. I never would've thought of that.

Al: (*Finishes writing. He steps back to look at the two lists side by side.*) Hmm . . . something's still missing. We haven't said exactly what our paper towel will do.

Figure 6.2

The two completed columns now include these items.

Goals	Things that limit goals
Strongest	people use for cleaning with scouring powder
Most attractive	must match typical kitchens
Most absorbent	average spill is about 8 ounces
Most convenient	size of a typical holder, strength of a typical tug
	materials

Explain

ISAAC: Sure we have! You just didn't write it down. For "Strongest," we said that when people use it with scouring powder, it won't shred or get holes in it.

MARIE: How do we do that?

ISAAC: Let's weave nylon thread into each towel!

ROS: Al, make another column. I think a good heading would be "Final decisions."

AL: Is your hand broken, Ros?

ROS: Okay, okay. (*Ros takes Al's place, draws a third column, and labels it "Final decisions." On the "Strongest" line, she writes, "Enough nylon thread woven into each towel so it won't tear when you use scouring powder."*) And I remember what we said for absorbent. What do you guys think of this? (*She writes, "Make the towel two thicknesses and large enough to wipe up a full glass of milk that spilled."*)

MARIE: Looks good. On the "Attractive" line write in "Blue, green, gold, brown, and flower patterns."

ISAAC: And on the "Convenient" line write, "Towels will tear easily at the perforated lines," and "The roll will be 11 inches wide."

AL: How do you know that, Isaac?

ISAAC: I just measured this roll of paper towels on my desk. Okay everyone, that's a wrap. Someone take that down to manufacturing quick so they can start making the perfect towel.
(*All of the characters stare icily at Isaac.*)

ISAAC: Well, I guess my legs aren't broken. I can do it myself.

Figure 6.3

The characters end up with a chart composed of three columns like this one.

Goals	Things that limit goals	Final decisions
Strongest	people use for cleaning with scouring powder	Enough nylon thread woven into each towel so it won't tear when you use scouring powder.
Most attractive	must match typical kitchens	Blue, green, gold, brown, and flower patterns
Most absorbent	average spill is about 8 ounces	Make the towel two thicknesses and large enough to wipe up a full glass of milk that spilled.
Most convenient	size of a typical holder, strength of a typical tug	Towels will tear easily at the perforated lines and the roll will be 11 inches wide.
	materials	

As designers, Al, Marie, Ros, and Isaac are right on target. By getting together and listening to one another, they have pointed out some important ideas about design that you need to understand before you proceed through this unit.

In order to design a product, people must first identify the goals for their product. Sometimes, but not always, people phrase the goals they set for their product using general words such as "most," "best," "largest," or "smallest." Or people can state their goals as specific properties that describe, such as "strong enough to pick up average spills." The characters' goals fit that pattern. The characters wanted their towel to be attractive, strong enough to scrub with, absorbent enough to handle a typical spill, and convenient to use. Designers, engineers, scientists, and manufacturers, among other professionals, call goals like these **criteria** ("criteria" is the plural for "criterion").

People judge a good product by whether or not it meets their criteria. For example, if the characters find that the towel they develop is not attractive to most people, they have failed to meet the "most attractive" criterion. Setting criteria is the first step in deciding on a product design.

In the characters' second column, there is a list of things that will limit or affect how they meet their goals. Often, people's preferences, lifestyles, and physical or mental limits have an effect on criteria. That means human factors play a big role in determining how manufacturers will meet the criteria. Remember, considering human factors helps produce a design that "fits" people. But as Rosalind pointed out, other things also can affect how designers meet their criteria, and these things in turn affect the final decisions. Most often these other things have to do with the manufacturer's budget, making a profit, and available or affordable materials.

Human factors and other things that affect the goals are called **constraints.** If designers tried to make the final design decisions based only on the criteria, they would reach a stopping point. At this point they would realize that they cannot make just any decision. They are limited by the materials they can use or by certain human factors of the consumer. So determining what constraints there are on the decisions is the next step in designing a product.

Only after you have decided on the criteria and have determined the constraints can you focus on specific ideas about what the product will do or what the product will look like. The last step in product design is making the final decisions based on the criteria and constraints. When you have gone through these steps, as the characters have, you can be fairly certain that your product will fit consumers.

 **CONNECTIONS:**
Do You Understand Criteria and Constraints?

When you design something as simple as a trash can, you go through the same steps as when you design something as complex as a computer. Meet with your team of three from the previous investigation to answer the following questions about the classroom trash can. After you agree on answers, record them in your notebook. Be prepared to share your answers in a class discussion.

1. List the criteria the manufacturers might have established for the trash can.
2. List the constraints that might have affected the design of the trash can.
3. List the decisions that the manufacturers made when designing the trash can in your classroom.

INVESTIGATION:
Part of Your Complete Breakfast

FAX MEMO

Date: Today, This Year
From: Your Editor
Re: Your latest writing assignment for *Teen Consumer's Magazine*

Dear Pulitzer-Prize winning writing team:

I am faxing this memo to you because it requires your immediate attention. This month's issue of *Teen Consumer's Magazine* goes to press in three days, and we need one last article. We want to include an article on breakfast cereals for our nutrition column. Please research this idea. You will need to test breakfast cereals and rank them from best-overall cereal to worst-overall cereal for the average teen consumer.

After you have completed your research and performed tests on the cereals, please write an article for our magazine explaining your process. Be sure to tell the readers what criteria you considered and evaluate the products according to how well the manufacturers accounted for constraints. I'd like you to use the same format as you did when you wrote that article for *Consumer Reports* on paper towels. Please fax your article back to me and include all diagrams and art work that you can.

In case you have lost yours, I have enclosed the step-by-step procedure that all reporters for *Teen Consumer's Magazine* are supposed to follow. I expect your article in three days.

Working Environment

Work cooperatively in your team of three. You will need the roles of Communicator and Manager. Practice the new social skill Speak softly so only your teammates can hear you. Move your desks together as you did in Tall, Dark, Handsome, Strong, and Absorbent.

Materials

For each team of three students:

- 1 measuring spoon
- small cups for the cereal samples
- a supply of water
- 1 tray
- any other materials your teacher provides that you decide you need for your tests
- 1 medicine dropper
- 3 craft sticks, toothpicks, or plastic spoons

Procedure: Part A—The Social Skill

1. With your teammates, decide why speaking softly so only your teammates can hear you is an important social skill.

   Notebook entry: Record two of your ideas.

2. Construct a T-chart for this social skill in your notebook.

Procedure: Part B—The Activity

1. Obtain the materials.
2. Make a list of the criteria that you feel are important for judging the quality of cereals.

   This should be one column in a data table.

3. Make a list of the constraints that might affect the design of cereals.

   This should be another column in a data table.

4. Decide on operational definitions to test your criteria.

   Notebook entry: Record the definitions.

5. Perform your tests and rate the cereals according to each of your criteria.

6. Construct a ratings guide table for the cereals you test.

   You can refer to the ratings guide in the Consumer Reports *article for ideas.*

   Notebook entry: Record your ratings guide.

Wrap Up

Compose an article for *Teen Consumer's Magazine.* Include information such as general comparisons among the cereals, what you tested and why, how you tested things, any art work or diagrams that you would like to put in the magazine, as well as your test results and cereal rankings. Refer to the article about paper towels as often as you like and have the Communicator get help from other teams. Present your article to the entire class and describe how successful your team was in using the unit skill and in trying to speak softly.

CONNECTIONS:
Evaluating Your Understanding of Criteria and Constraints

Study the character situations on the following pages. Discuss them in your team of three. Construct a table in your notebook with a column labeled "Criteria" on the left side, a column labeled "Constraints" in the middle, and a column labeled "Decisions" on the right side. For each product that the characters are discussing, list all of the criteria you think the manufacturers had in mind for that product. Then list the constraints you think limited the manufacturers' decisions and the final decisions the manufacturers made. When you are finished, put your name on your paper and give it to your teacher.

Situation #1. The meal tray

Figure 6.4

Study this top view of what their meal trays look like as you determine the criteria, constraints, and final decisions of the designers.

Situation #2. The notebook

Figure 6.5

This is the type of notebook Al has. Study it carefully to determine the criteria, constraints, and final decisions of the designers.

Situation #3. Marie's headache

Figure 6.6
The inside of an empty locker at Ros's school looks like this. Study the features carefully to determine the criteria, constraints, and final decisions of the designers.

Evaluate

Consumer Concerns ■ 129

CHAPTER 7

Your Designing Ways

Now that you have discovered three important parts of designing products (can you name them?), you are ready to explore the design process itself. How do designers make their decisions? What steps do they take in order to create a product that meets their needs? What are their criteria and constraints? In this chapter you will have an opportunity to design products using the steps that professional designers use. You also will learn more about two fascinating topics: boats and children.

INVESTIGATION:
Bon Voyage, Tom Thumb!

According to an old legend, Tom Thumb was a tiny man, no bigger than an average man's thumb. In the first part of this chapter, you will design and build boats, ones that perhaps Tom Thumb himself could have used in his travels. But before you do that, it will help if you know something about how boats move through the water. This investigation and the one that follows will help you explore boat propulsion.

Materials

For the entire class:
- several tubs or sinks for holding water
- a water source
- Tom Thumb boat fuel in a small cup
- 1 medicine dropper

For each team of three students:
- 1 pair of scissors
- 1 sheet of unlined, white paper
- 1 metric ruler
- 10 medicine droppers

Working Environment: Work cooperatively in your team of three, moving your desks together or sitting together at a table. Practice the social skill Speak softly so only your teammates can hear you. Use the roles of Communicator and Manager as well as Team Member roles.

Procedure

1. Watch closely as your teacher performs a demonstration.

   Notebook entry: Record any observations you make during the demonstration.

2. Obtain the materials for your team.

3. Design your version of a Tom Thumb boat.

   You have only five minutes in which to design your boat. Make your boat as similar to or as different from your teacher's as you would like, but you will have to use the same fuel to propel your boat that your teacher used. Your boat also should perform as well as or better than your teacher's boat. Do not put your boat in the water yet.

   STOP: Use this opportunity to be open to the ideas of both your teammates.

4. Measure your boat.

   Notebook entry: Record its length and width.

5. Prepare a data table in which you will record the size, a description of the shape, a description of the direction traveled, and the general speed of all the boats in the class.

   Ask your teacher how you will determine the speed each boat travels.

6. Demonstrate your boat to the rest of the class by putting your boat in the water and adding the fuel.

7. Observe what happens to your boat, as well as what happens to the other students' boats.

Notebook entry: Record your observations in the appropriate columns in your data table. Tell the class your boat's measurements.

8. As a team decide which boat from all those in the class was the best based on speed and direction.

9. Compose a statement that tells what type of Tom Thumb boat goes the fastest and what kind of Tom Thumb boat goes the straightest.

Notebook entry: Record your statement.

Wrap Up

Discuss the following questions as a team. You then will discuss them as a class, so be sure each of you can answer each question.

1. What type of Tom Thumb boat goes the fastest and straightest? Does the same type of boat that goes the fastest also go the straightest?

2. What were some common criteria and constraints that each team had while designing and making a Tom Thumb boat?

3. What is Tom Thumb fuel, and why does it propel a Tom Thumb boat? It's okay if you don't know for sure. Just try to explain it as well as you can.

4. Rate your team on a scale of 1 to 10, 10 being the highest, for how well you used the unit skill. Then rate yourself, using the same scale, for how well you practiced the skill of speaking softly so that only your teammates can hear you.

READING:
Is a Boat a Boat?

A boat is a boat is a boat, right? "No," you say? Well, if you disagree that a boat is a boat, you are right in a certain respect. There are different types of boats: sailboats, speedboats, yachts, cruise ships, battleships, canoes, rafts, motorboats, houseboats, steamboats, tugboats, freighters, rowboats, paddle boats, aircraft carriers, garbage scows, junks, oil tankers, sloops, bilanders, rub-a-dub-dub-three-men-in-a-tub boats, barques, schooners, catamarans, gondolas, sampans, sculls, sharpies, kayaks, barges, Yankee clippers, ferries, dinghies, dories, dugouts, skiffs, ketches, punts, feluccas, outriggers, galleons—get the picture? If you agree that a boat is a boat, you are also right. There may be different types of boats, but all boats are basically the same.

Okay, how can a raft be like an oil tanker? How can a garbage scow be like a cruise ship? How can a paddle boat be like a speedboat? When you put images of these pairs in your mind, you probably see two vessels that look and act very differently. Imagine the slow, somewhat clumsy crawl of a paddle boat next to the sleek, slim figure of a speedboat cutting through the water at top speed.

Or think of a garbage scow reeking of fumes in a busy harbor and desperately searching for a place to unload its months' worth of waste next to the glamorous, white silhouette of a cruise ship gliding along crystalline waters. These boats are not the least bit the same! Well, no, not in appearance, anyway. But they share these three criteria:

- First, a boat has to float.
- Second, a boat must stay level and not tip over in the water.
- Finally, a boat must have some means by which it moves through the water.

If manufacturers did not meet these criteria when designing boats, then we wouldn't have any boats at all! These three criteria make them boats.

Once boat builders meet these common criteria, then other criteria as well as constraints, or limits, make each boat a unique vessel. For example, a boat builder decides to make a boat using the following criteria:

- the boat floats,
- the boat is engine-powered,
- the boat will not tip over in the water when it is used under normal conditions, and
- the boat will provide recreation for people.

The constraints that will limit this boat builder's decisions are the following:

- the materials he or she will use,
- the fact that most engines require a fuel-powered motor,
- the fact that many people define recreation as relaxation and fun, and
- the depth of the water in which the boat typically will be used.

The final decisions a boat builder makes, then, are that the boat will be made of steel, the motor will use diesel fuel, and the boat will be similar to a hotel with recreational facilities. This boat builder, then, has decided to build a cruise ship.

In the previous investigation, Bon Voyage, Tom Thumb, you were boat designers working with specific constraints and criteria. You were told to make a Tom Thumb boat with this criterion:

- the boat was to perform as well as or better than the teacher's boat.

You were given these constraints:

- the boat was to be propelled by the fuel the teacher used,
- you needed to use the same materials the teacher used, and
- you could take only 5 minutes to design and build your boat.

SIDELIGHT

This Wonderful World of Boats

Technology involves people solving problems. One of the oldest human problems is how to get people and things from one place to another—the problem of transportation. Using boats as a means of transportation is an ancient idea. In fact, boats are one of the oldest ways of transporting things and people. The only other type of transportation that came before boats, besides walking, was the use of sleds for heavy cargo. Between 5000 and 3500 B.C. (that is between 7,000 to 5,500 years ago), people began using donkeys and oxen to pull sleds of cargo. It was also at this time that people first built boats in the form of simple rafts. Later people learned to build canoes and dugouts.

The Smithsonian Institution Photo #38807

that time people had invented the wheel. The invention of the sailboat and wheel are two technologies that helped people the most to overcome obstacles in transportation.

The Bettmann Archive

People used these early boats as a means of transportation along rivers, streams, and lakes. This greatly increased the contact among people and advanced the trading of goods. It wasn't until people invented the sailboat in 3200 B.C. that transportation across the ocean was possible. By

The Smithsonian Institution Photo #58589

Try, if you can, to imagine a world without boats. What are some things that might be different in your life if it weren't for boats?

Explore   Your Designing Ways ■ **135**

You didn't have to worry about whether or not your boat would float because you already had seen that your materials would float. You didn't have to worry about whether or not your boat would tip over because the boat was flat and carried no cargo. You did, however, experiment with propulsion, or what made the boat move. Some boats have sails and move using the wind, some boats have steam engines, some have coal-burning engines, and some have electric engines. Some boats are powered by humans using poles, oars, or paddles. What you probably discovered in your experiments is that the size and the shape of boat has an effect on how the boat moves.

What if someone asked you to design and build a boat that floated, did not tip over, and was propelled somehow? Could you do it? What if you also had to decide what job your boat would perform: Would it carry cargo or passengers? Would it be for recreation or work? Guess what? That's exactly what you will do next in this chapter! First, however, you will need to practice with different methods of propulsion.

INVESTIGATION:
Sails, Propellers, and Gas

In this investigation you will experiment with three major forms of boat propulsion. When you are finished, you should know several ways to move objects across water without pushing them yourself. You will build boats that are more complex than the Tom Thumb boats but not as refined as real boats. You will use what you learn in this investigation to design and build the best boat possible in the next investigation.

Working Environment

Work cooperatively in your same team of three. Continue to practice the social skill Speak softly so only your teammates can hear you. Use the roles of Communicator, Tracker, and Manager. Create a large work area with your desks or at a lab table. You also will need a test area that is a tub or sink full of water.

Materials—Parts A, B, and C

For each team of three students:
- one 2.5-by-10-by-15-cm block of wood with a small hole drilled through it
- 1 stick of modeling clay
- access to a tub or sink full of water
- 1 metric ruler

Materials—Part A

For each team of three students:
- 1 toothpick, craft stick, or new, sharpened pencil
- 3 sheets of white, unlined paper

Materials—Part B

For the entire class:
- 1 roll of transparent tape

For each team of three students:
- 1 rubber band, 4-by-$\frac{1}{4}$ in.
- 2 paper clips
- 1 plastic bead
- 3 push pins
- 1 paper propeller you choose from the Propellers sheet
- 1 aluminum foil pie plate or baking pan
- 1 pair of scissors
- strips of electrical tape (get these strips when you come to the step in which you need them)
- 3 pairs of goggles

Materials—Part C

For the entire class:
- 1 roll of masking tape

For each team of three students:
- 1 round balloon, 10 in.
- 2 Alka-Seltzer™ tablets
- a water source
- 1 flexible straw
- 1 squeeze bottle, 250 mL

Procedure: Part A—Sails

1. Obtain all of the materials for Part A.

   This includes materials that are common to Parts A, B, and C.

2. Make a sail for your wood block by piercing two holes through a piece of paper with a toothpick, craft stick, or pencil.

See Figure 7.1. The toothpick, craft stick, or pencil is called the mast.

3. Attach your sail to the base of your block with a piece of modeling clay.

   See Figure 7.2.

4. Create a wind that is strong enough to sail your boat across a tub or sink of water.

   Be sure to speak softly so only your teammates can hear you if you share a sink or tub test area with other groups.

Figure 7.1

This is an example of what a mast and sail look like. Make sure the pencil point is down.

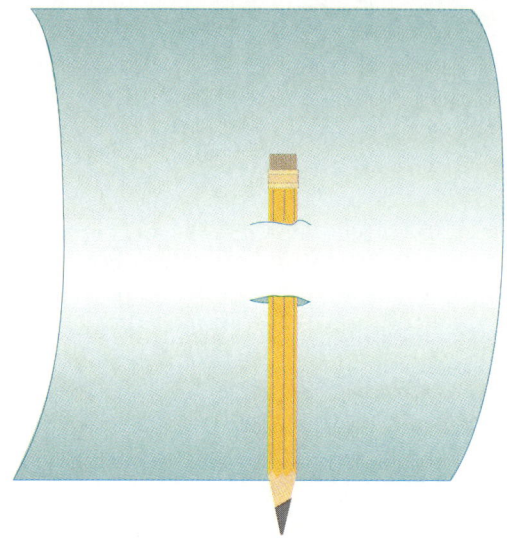

Figure 7.2

This is one way to attach your sail and mast to your block using modeling clay. Make sure the pencil point is down into the lump of clay.

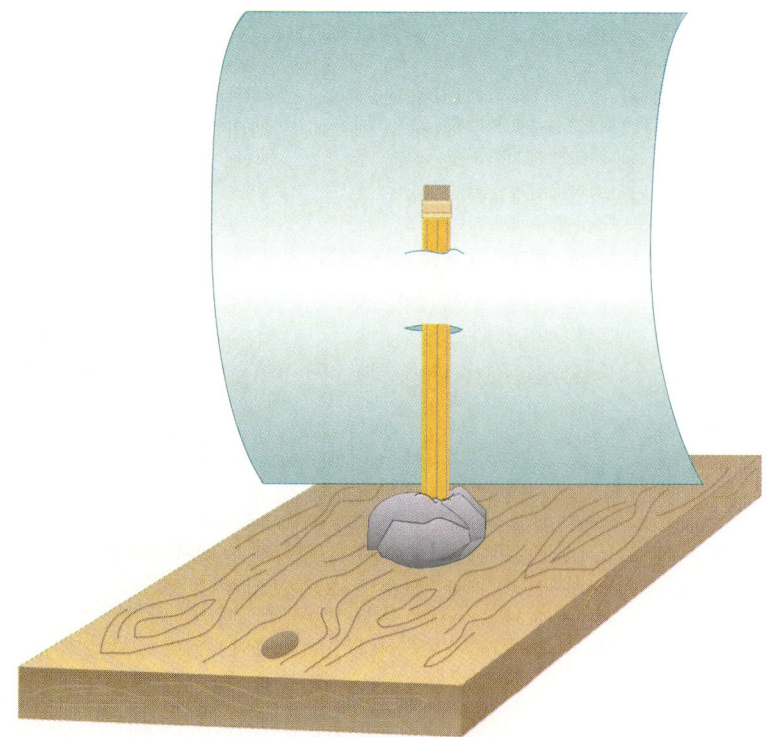

5. Experiment with different sizes and shapes of paper, different masts, and the placement of the mast on the block to make the most effective sail for your boat.

   Be sure to identify and control your variables as you experiment with the materials. Remember to be open to others' ideas.

6. Construct a data table to record your attempts: paper sizes and shapes, mast sizes, and mast positions.

   You should include a space for recording what you observe with each change in sail, mast size, and placement of the mast.

7. Record your data in your data table.

   Record every sail size, sail shape, mast size, and mast position that you try.

8. Put a star beside the type of sail that worked the best.
9. Return all the Part A materials, but keep the materials common to all parts.

Procedure: Part B—Propellers

▲ **CAUTION:** For this procedure you will need to wear eye protection, such as goggles.

1. Obtain all of the materials for Part B.
2. Trace a paper propeller onto an aluminum foil pie plate or baking pan, and cut out the propeller.

   Choose any propeller pattern.

▲ **CAUTION:** Propellers cut from aluminum foil pans might have sharp edges. Be careful not to cut yourself.

3. Set the push pins into the wood block as in Figure 7.3.

   Be sure the push pins are as far down into the block as they can be. If they are not, they can fly into the air later when you attach the rubber band.

Figure 7.3

The two push pins are as close to the edge of the block and as close to each other as possible. The single push pin is about 6 inches from the two push pins. Place the push pins in these positions along the center of the block. Squeeze the two push pins together if you need to.

Explore

Figure 7.4

(a) Unbend the paper clip. Notice that one end is smaller than the other. (b) Bend the paper clip up at the small end. The small end of the clip should form a right angle to the large end of the clip. (c) Slip the bead onto the smaller end of the paper clip until it stops at the kink. (d) Bend the smaller end back into a hook.

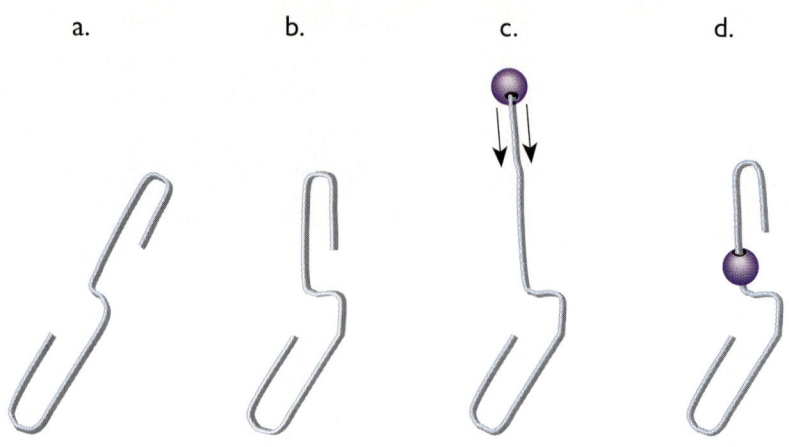

a. Unbend the clip. Notice that one end is smaller than the other.
b. Bend the paper clip up at the small end.
c. Slip the bead onto the small end of the paper clip until it stops at the kink.
d. Bend the small end back into a hook.

4. Follow the steps in Figure 7.4a through 7.4d to prepare the paper clip and bead assembly.
5. Hook the small end of the paper clip through the rubber band.

   Squeeze the small end of the paper clip to keep the rubber band from slipping off.

6. Attach the paper clip assembly to the board as in Figure 7.5.

   Be sure you are wearing eye protection!

7. Twist the rubber band by winding the large end of the paper clip 30 to 35 times.
8. Release your hold on the large end of the clip to see if it spins back like a propeller.

   If it spins like a propeller, proceed to step 9. If not, adjust the pin, bead, or rubber band until you make it spin.

Figure 7.5

Attach the large end hook of the paper clip onto a rubber band stretched from the single push pin. Squeeze the hook shut. Place the bead on the opposite side of the two push pins.

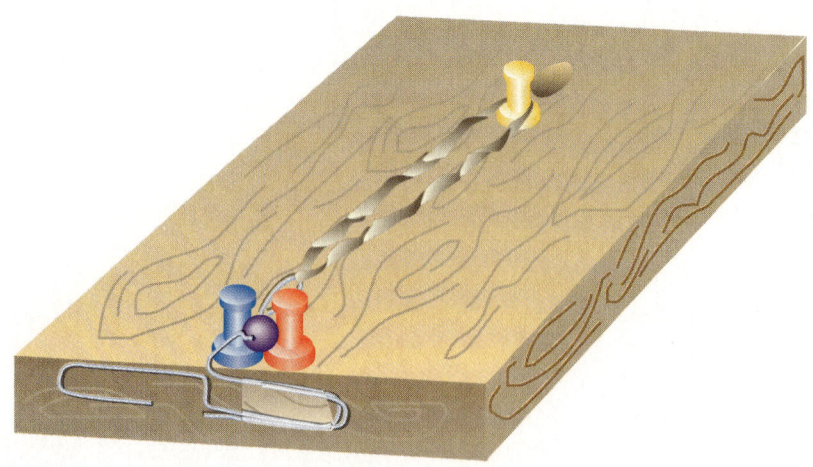

Figure 7.6

Tape the small end of the new paper clip to the small end of the clip holding the bead. Try not to use too much tape.

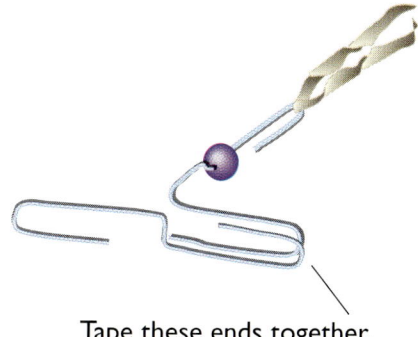

Tape these ends together.

9. Tape another unbent paper clip to the small end of the paper clip as shown in Figure 7.6.

   Use electrical tape.

10. Tape your aluminum propeller to the paper clip.

    Use electrical tape. Match and bend the aluminum propeller on the paper clip to achieve the best fit, as in Figure 7.7.

11. Gently twist the blades in opposite directions to shape the propeller as shown by the arrows in Figure 7.7.

12. Wind the propeller and release it once you have placed the wood block in a tub or sink full of water.

13. Observe what happens.

    Notebook entry: Record your observations.

14. Experiment with the different propellers to see which one provides the most propulsion.

    Remember to control variables as you try to determine what type of propeller works best.

Figure 7.7

Experiment with the shape and twist (or pitch) of the propeller. Try different amounts of pitch and different directions of pitch.

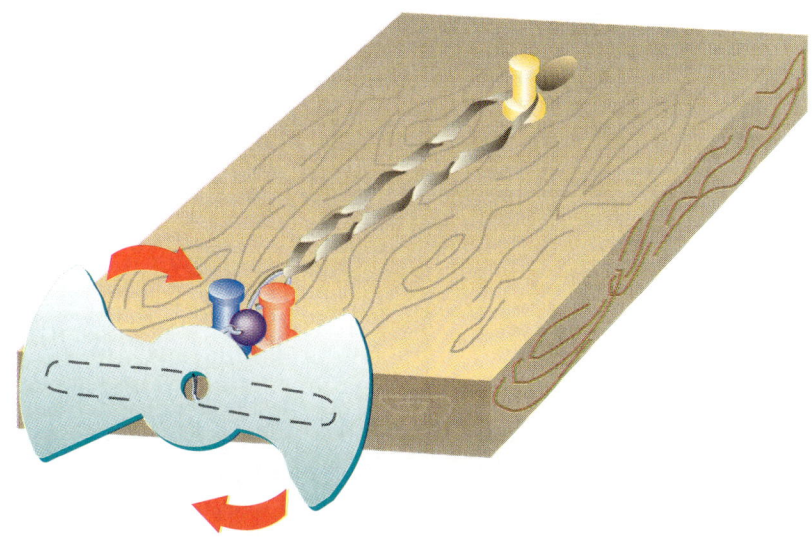

Explore

Your Designing Ways ■ 141

15. Construct a data table to record your information.

    Your data table should include the type of propeller, how many times you wound the propeller, how many twists of the blade make one full turn, and how you shaped the propeller blades.

16. Return the materials for Part B, but keep the materials common to Part C.

Procedure: Part C—Gas

1. Obtain the materials for Part C.
2. Blow up the balloon, then let the air out.

    The manager should do this and check to be sure there are no holes in the balloon. If there are any holes or leaks, obtain a new balloon.

3. Break each of the Alka-Seltzer™ tablets into four pieces.

    Have the Tracker do this.

4. Stuff all eight Alka-Seltzer™ pieces completely into the balloon.

    The Tracker should do this.

5. Fit the neck of the balloon over the end of the flexible straw that is nearest the joint.

    Have the Communicator do this.

6. Wrap tape around the straw and the neck of the balloon to seal the balloon against the straw.

    While the Communicator holds the neck of the balloon tightly in place around the straw, the Manager should wrap the straw and neck of the balloon six times with tape to make a tight seal between the balloon and the straw. See Figure 7.8.

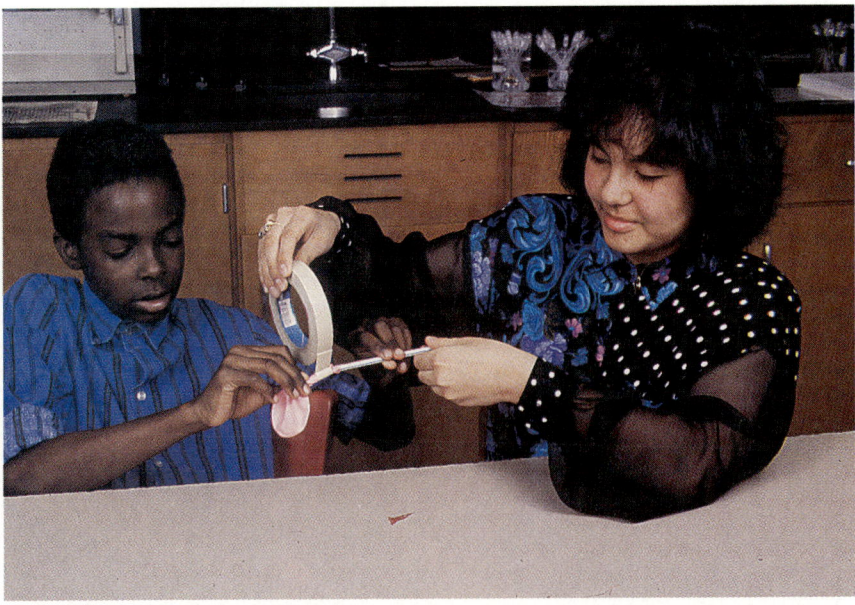

Figure 7.8

Create a seal between the balloon and the straw by wrapping masking tape at least six times around the edge of the balloon's neck and the straw.

7. Blow through the opposite end of the straw to inflate the balloon half-full.

   As the Tracker blows, the other Team Members should help him or her listen for air leaking out of the seal made by the masking tape. If the Tracker can't blow up the balloon, and air is leaking out of the seal, the Manager should squeeze the tape tighter around the balloon and straw or should apply more tape to make a better seal. It is important to make sure that there is no air leaking out of the straw and balloon.

8. Let the balloon deflate.

9. Place the open end of the straw through the hole in the block, from the top of the block to the underside.

   Have the Tracker do this. The balloon should be on top of the block.

10. Bend the straw so that the tail of the straw is parallel to the underside of the block.

    The Tracker should do this. The flexible part of the straw should be in the hole.

11. Wrap a thin rubber band around the block and straw to hold the tail of the straw against the underside of the block.

    Have the Communicator do this. Be sure the straw is not kinked in any place, especially at the flexible part.

12. Cut off about 8 cm from the tail of the straw.

    Have the Manager do this. The entire apparatus should look like the picture in Figure 7.9.

13. Fill the 250-mL squeeze bottle with water.

    Have the Communicator do this.

14. Squeeze water into the tail of the straw until the main part of the balloon is full of water.

    While the Tracker turns the block down, holding it so the tail of the straw is pointing up, the Communicator should vigorously and quickly squeeze the water into the main part of the balloon. They should do this until the balloon is full of water. If you add the water too slowly, the gas will form and dissipate before you are finished

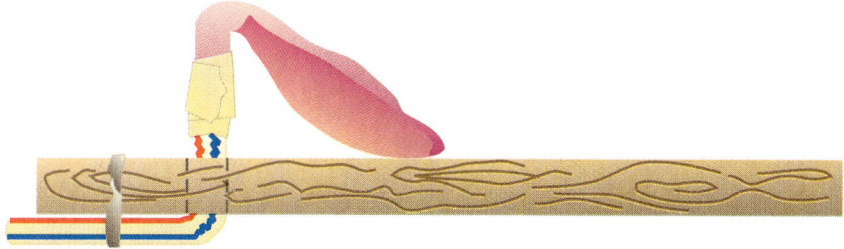

Side View

Figure 7.9

This is what your gas apparatus should look like now. Notice how the straw bends in the hole of the block.

adding water. You will then have no gas left to propel your block. See Figure 7.10.

STOP: Are you speaking softly so only your teammates can hear you?

15. Remove the squeeze bottle from the straw, and immediately seal the open end of the straw with your index finger.

    As the Communicator removes the bottle, the Manager should cover the end of the straw with his or her finger as quickly as possible. Try not to let too much gas escape.

16. Lower the block into the water, balloon side up.

    The Manager should keep his or her finger covering the end of the straw to prevent any air from escaping.

17. Wait until the balloon inflates one-third to one-half full and then uncover the end of the straw.

    If the balloon does not inflate with gas, you probably have a leak between the balloon and the straw. You will need to begin Part C over again if this is the case.

18. Observe what happens to the block.

    Estimate how far the block traveled before stopping and how many seconds it took to stop.

    Notebook entry: Record these observations.

19. Discuss with your teammates how you could make the block go farther or faster.

20. Experiment with any of the materials from this investigation to design a better gas engine.

    Be sure to control variables to ensure a fair test each time.

Figure 7.10

Do this step *quickly*. You might not have time to fill the balloon completely with water if you squeeze the water in too slowly.

21. Record the observations and results of your tests in a data table.

   Your data table should include information about the container you used for the engine, how many tablets you used, how much water you used, how far the block traveled, and how fast the block traveled.

22. Draw a star beside your entry for the engine that gives you the best results.

23. Return all of the materials to their appropriate place.

Wrap Up

As a team decide on answers to the following questions and be prepared to share your answers with the rest of the class. Record your answers in your notebook.

1. Which method of propulsion worked the best for your team? Why?
2. Explain how easy or difficult it was for your team to speak softly during this investigation.
3. What specific strategies can you try in the next investigation to improve your use of the skill of speaking softly so that only your teammates can hear you.

INVESTIGATION: Anchors Away!

In this investigation you will have a chance to use what you know about boats from Bon Voyage, Tom Thumb, what you learned in the reading Is a Boat a Boat, and what you discovered in Sails, Propellers, and Gas. You will try to build a boat that meets specific criteria and accounts for specific constraints.

Materials

For each team of three students:
- Anything from the materials provided by your teacher. You may not use any other materials.

Procedure

1. Read the following challenge:

   Build a boat that floats, does not tip over in the water, has a means of propulsion other than you pushing it, and meets one specific purpose. Your teacher has priced the materials you will use and will tell you how much you are allowed to spend on your materials.

2. As a team decide what kind of boat you want to build.

Working Environment

Work cooperatively in your team of three. You will need a Manager and a Communicator. In this investigation concentrate on the skill Be open to others' ideas. You again will need a large work space, and you will share a sink or tub with other teams. When you are with members of another team, try to use their names.

For example, decide whether you will build a ferry boat, cargo boat, cruise ship, speedboat, or any other kind of boat you can think of.

3. Construct a data table with three columns. Label the first column "Criteria," the second "Constraints," and the third "Decisions."

Fill in this data table with appropriate entries as you design your boat.

5. Test your boat in a tub or sink full of water.

Check your data table to be certain that the boat functions according to your goals.

6. Redesign and retest your boat until you are satisfied that your boat is the best it can be.

7. Put away the materials.

Wrap Up

Complete questions 1 and 2 with your team, and prepare a presentation as described in question 3. Make sure that each of you can explain your answers and that each of you takes part in the presentation.

1. Point out to your teammates at least one specific instance in which you noticed how someone was being open to others' ideas.

2. Discuss why concentrating on the unit skill was important for this investigation.

3. Present your criteria, constraints, and decisions table to the rest of the class, and show them your boat in action. Ask the class for suggestions that would make your boat even better.

CONNECTIONS: Technological Problem Solving

At this moment you are probably surrounded with products of modern technology. You are probably at a desk, seated on a chair, with some sort of writing implement handy, and you most likely have your notebook beside you. The classroom itself might be furnished with an overhead projector, a chalkboard, lab space with sinks, laboratory equipment racks, cabinets, shelves for books, a trash can, a clock, and perhaps a bulletin board.

These items are just a few of the products in your everyday life that you might take for granted. Until the last investigation, perhaps you hadn't thought too much about boats, either. Yet all of these products went through a process of design similar to the process your team used. Engineers and other kinds of designers use a design process (as you did) every time they design a new object. In fact, all of the things you see around you in your classroom probably required hours of planning and engineering before the designers felt that they had met their product goals.

"Wait a minute!" you might feel like saying, "What design process did we use?" Well, believe it or not, your team developed your own design process in the investigation Anchors Away! You might not have been aware of it, so this connections section will help you identify the unique steps of your team's design process. Work on this section with your teammates.

Cards

- With your teammates sort through the deck of Design Process Cards from your teacher.
- Discuss the meaning of each card so everyone understands what it says.
- Select the cards that fit the steps you used when you designed your boat. (Which card seems to say best what you did first, second, third, and so on? Which cards did you repeat?)
- Remove cards you didn't use. Use blank cards to write in any steps that you don't find among the written cards or to recopy a step you repeated.
- Create a new deck by reassembling in order the cards that you used and the new cards you wrote. Save any spare cards in a separate pile.

Flow Charts

- Arrange the cards in a pattern (like a chart, map, or diagram) that shows the order of the steps in the process your team used in designing your boat.
- Tape the cards into place on a poster board or large sheet of paper.
- Be creative in the pattern in which you arrange the cards.

Presentations

- Present and explain your finished chart to the rest of the class.
- Listen to other teams explain their own design processes. Ask questions if you don't understand something.
- Other teams' designs might remind you of a step or two that you did but forgot about. After listening to other teams, you might then want to change your design process chart by rearranging the cards on your chart or by adding cards from your spare cards pile.

An Example of One Design Process

- Copy your team's design process chart into your notebook.
- Ask your teacher where you can store your poster.
- Review and discuss the characters' design process that follows.

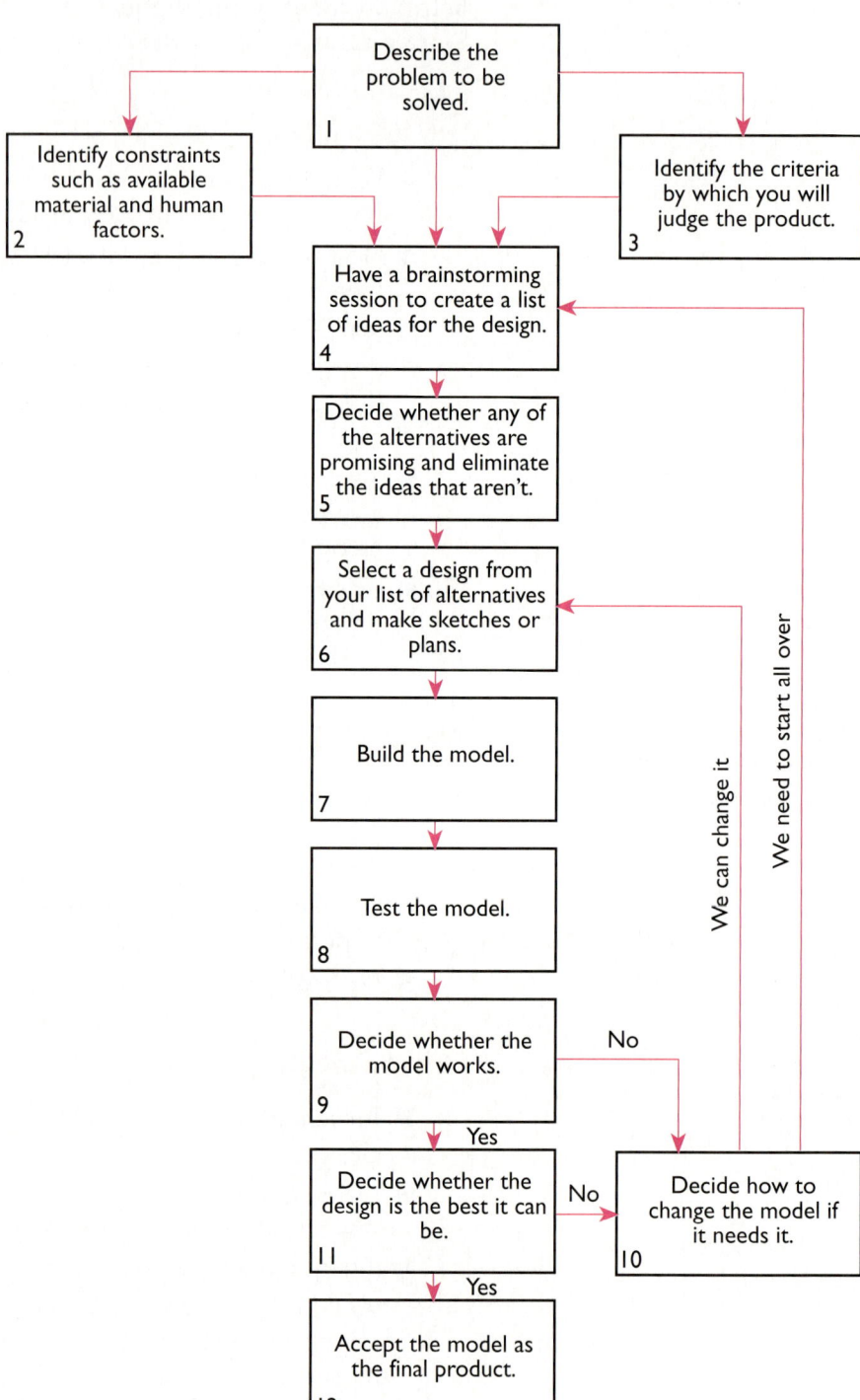

Figure 7.11

This is the way the characters organized their cards into a design process. They felt that this best described the process they used when they designed their boats in Anchors Away!

Figure 7.12 This was the characters' first step.

Figure 7.13
The characters combined two cards to describe what they did next.

Explain Your Designing Ways 149

Figure 7.14

The characters conducted a brainstorming session that helped them fill in their decisions column.

5 Decide whether any of the alternatives are promising and eliminate the ideas that aren't.

6 Select a design from your list of alternatives and make sketches or plans.

Figure 7.15

The characters talked through their plans and drew some sketches. This helped save materials and kept them within their budget.

7 Build the model.

8 Test the model.

Figure 7.16

After their first model was complete, the characters tested it in water.

Explain Your Designing Ways 151

Figure 7.17
The characters' model sunk, so Marie suggested ways to change their basic design.

Figure 7.18
The characters decide that their design is the best and decide to stop changing it.

Making Connections

Answer these questions in your notebook.

1. In Chapter 6 the characters portrayed a scene in which they were paper towel designers. What step in that scene did the characters reach in their own design process?
2. What step in your design process did they reach?
3. Why didn't the characters finish their process?
4. Now look back to Unit 1 when you were designing the ultimate TV. Decide which of the steps in your flow chart you used when you designed it.

You and your team went through some sort of design process just as Marie, Al, Isaac, and Ros did. As you discovered, and the characters vividly illustrated, design is not just a haphazard series of events. Design depends on people working together most of the time and following a process for technological problem solving the way your team did and the way the characters did.

Actually, the design process chart you created, also known as a flow chart, represents how designers might go about creating products. Most designers, however, would not use a chart, but a chart would describe the process they use in designing. Often a design process is as simple as thinking, creating, testing, and modifying. Taking the time to follow a design process means that it will take less time in the long run to design a product, particularly if you also know how to cooperate and work together as a group.

We've used the word "technological" quite a bit. Just what is **technology**? It is more than just products. It is also the process that one goes through to develop those products. It is the process you just went through as design engineers.

INVESTIGATION:
Toys for Tots

Now that you have created your own flow chart for technological problem solving (your design process chart), you are ready to use it to design something else. In this investigation you will have a design opportunity that might come only once in your lifetime: a chance to design a toy.

Materials

For each team of three students:

- Any materials that you choose from those provided by your teacher. You may not use any additional kinds of materials.

Procedure: Part A—The Task

1. Read the following challenge.

   Use your team's process for technological problem solving from the previous connections section to design and construct a toy. The toy should meet the following criteria:

   - Be interesting to a child who is between 3 and 8 years of age.
   - Be safe to use.

   See Figure 7.19. You can revise your design process chart any time as you design your toy. Make changes on your chart in your notebook.

Working Environment

Work cooperatively in your team of three. Use the roles of Communicator and Manager. Try to be the team that speaks most quietly to one another as you work. At the end of this investigation, your teacher will let you know which team was the most quiet. Gather with your team at a table or at your desks. You will need plenty of work space.

A safe toy is one that won't cause injury when a child plays with it. A designer has made a safe toy if the toy meets the following standards:

 It has no small parts that young children could swallow easily, inhale, or choke on. (Do not use balloons or parts of balloons in your toy design.)

 It has no strings that are 12 inches or longer and has no strings that together can form a circle 14 inches around.

 It has no sharp points on edges (e.g., protruding nails, metal edges, or glass).

 It contains no toxic paint or other chemicals.

 It contains no flammable materials.

Figure 7.19

In order to pass the "safe to use" criterion, your toy must meet all the points on this checklist.

2. With your teacher and the rest of the class, discuss the following:
   a. What are two constraints your team has for designing a toy?
   b. How could you determine the human factors of children 3 to 8 years of age that will affect the design of your toy?
3. As a class develop a chart that every team will use to record the human factors that will affect the toy design.
4. In your teams construct the human factors chart in your notebooks.

   Make the chart neat and leave enough room to record any additional human factors that will affect your design.
5. As a class operationally define "interesting to a child."

   That is, decide how the teams in your class will measure whether or not a toy is interesting to a child.

Procedure: Part B—The Process

1. Choose an age group for which your team will design a toy.

   It can be for any age from 3 through 8. Pick a group that you will not have too much difficulty researching.
2. Devise a plan to conduct research on the age group you chose.

   Are each of you also assuming the role of Team Member?
3. Have the Communicator check your plan with the teacher by informing him or her of the age group you chose and how you will collect information from that group.
4. Conduct a brainstorming session to determine a list of things children in the age group you chose might find interesting.

   Would your age group prefer a game, a doll, or other types of toys?
5. Complete the human factors chart. Follow these steps to determine human factors for the age group you chose:
   a. Decide on the gender, size, and any other characteristics of the children who are going to play with your toy.
   b. Make a list of the human factors for which you want to gather information.

      Some possible factors to consider are hand and finger size, strength, and general color preference.
   c. Select several children who will provide you with the information that you need.

      The children could be brothers, sisters, cousins, or friends of one of the members of your team. If you have an elementary or preschool nearby, you might gather data from the students there if your teacher has set up this opportunity for you. Make sure that you obtain permission from the parents, guardians, or teachers of the child or children that you plan to interview and from whom you obtain physical measurements.

   d. Only gather information for the human factors that affect your design project.

   Remember, they are doing you a favor!

   e. Keep the identity of the people you interview a secret. Tell them that they will remain anonymous.

6. Construct a data table that lists the criteria, constraints, and decisions that will be important in the design of your toy.

7. Fill in the columns in the table.

   Do this as you design your toy according to your own design process. Remember to adhere to the criteria, constraints, and human factors you have listed previously. Also remember that the third criterion is now limited by your operational definition of "interesting." Finally, you may use only the materials that your teacher provides. If you would like to work at home, take home the materials you chose to use or use similar materials from home.

8. Review your toy with your teacher to make sure that it passes all the safety criteria.

9. Return to the group of children who helped you determine the human factors for your chart and have them test your toy to make sure that its design has fulfilled all of the criteria.

   Before allowing any children to play with your toy, be sure it has passed a safety inspection in your team and then a safety inspection by your teacher.

10. If your toy does not fulfill all of the criteria, return to your design process and modify the toy until you are confident that it is better.

    Each time you modify your toy, you will need to return and test the toy again with the same children as before.

11. After your toy meets all of the criteria, find a team that is at the same step in this investigation and proceed to Part C.

Procedure: Part C—Does It Meet the Constraints?

1. Trade toys and charts of human factors with another team.

2. Manipulate or play with the other team's toy in order to determine whether or not the toy accounts for all of the human factors the team listed on its chart.

3. Meet with the team who evaluated your toy and whose toy you just evaluated and discuss your toys.

   Take turns discussing the good qualities of each other's toy and in a respectful manner point out any human factors in the chart that you did not think the other team accounted for. Help each other figure out ways to account for these human factors.

4. Present your toy to the rest of the class.

Wrap Up

Discuss the following with your team. After agreeing on answers, write the answers in your notebooks.

1. Explain any changes you made in your team's design process chart.
2. Copy a neat edition of your new chart on a clean page in your notebook.
3. Use your own rating or grading system to evaluate your progress at using the skill of speaking softly so that only your teammates can hear you.

SIDELIGHT

Ergonomics—The Science of Human Factors

Ergonomics is the branch of science that studies how people interact with the products, equipment, environments, facilities, and procedures they use at work and in their homes. Ergonomics specialists apply information about human factors—capabilities, limits, characteristics, and behaviors—to the design process to make sure that things are safe, comfortable, and easy for people to use.

Ergonomics specialists work in many different industries, including the aerospace industry, computer industry, communications, and consumer product industries. They also work in various businesses, such as general research and development businesses, health care, management consulting, and architectural services. Ergonomics specialists have bachelor's, master's, or doctoral degrees in a variety of subjects, including psychology, engineering, computer science, and industrial design.

Designing even commonplace products, like your toothbrush, can involve ergonomics. Not long ago a major manufacturer hired an ergonomics company to design an improved toothbrush. Specialists at the ergonomics company discovered that no human factors research had ever been used in toothbrush design, so they began an extensive study. First they read about dental care and interviewed dentists. Next, they distributed a questionnaire concerning dental care habits to 300 adults and analyzed the information they received. They then began collecting measurements of hands, teeth, and mouths of consumers. To find out how users handle a toothbrush, the design team studied films of people brushing their teeth. They looked at how people held and moved the brush and how much time people spent brushing different parts of their mouths. From these studies one criterion the design team determined was that the toothbrush handle should make the brush easy to maneuver. A constraint they identified for the design of the handle was that it couldn't be too wide to fit into the standard bathroom toothbrush holder.

After they examined the results of laboratory studies pertaining to plaque removal, the team designed two prototype (model) toothbrushes and manufactured enough for testing. Test subjects (the people testing the toothbrushes) compared the two prototypes to two common toothbrushes that were available already. Both prototypes removed plaque better. But the test subjects preferred the bristle head of one prototype and the handle design of the other prototype. So, the design team combined the features into a single product—the Reach™ toothbrush.

CONNECTIONS:
Human Factors as a Design Constraint

Work on this section in your cooperative team of three.

You now have had a chance to design toys for young children and to evaluate how well other teams designed toys. As a class you listed things that were important to consider when designing and building this toy. No doubt many of the following human factors are on your list:

- Physical characteristics—eye to hand coordination, height, hand size, and muscle strength of children of different ages. These physical **constraints** limit the ability of a child to manipulate and play with the features of a toy. A constraint is something that sets limits on what you can do.

- Mental characteristics—reading level, math skills, and logical-thinking skills of children of different ages. These mental constraints limit the ability of a child to understand and figure out the operation of a toy or game.

- Behavioral characteristics—for example, small children are likely to put objects in their mouths. These behavioral constraints limit the ability of children to play safely with a toy.

Physical, mental, and behavioral characteristics are usually the three types of human factors that design engineers try to account for as they design products. Each of these human factors sets a limit on what the designers can do. Your teacher will tell you how you will work to accomplish the following two tasks.

1. Break down your list of human factors into three separate lists: physical, mental, and behavioral. Write the lists in your notebook.

2. Compose a paragraph that summarizes what you learned in this chapter and what this chapter was all about. Write the paragraph in your notebook and be prepared to share your paragraph with the class.

SIDELIGHT

Lego™ Success Story

In the United States, nearly two-thirds of all homes with children under age 15 have a Lego™ set. In Europe, the figure is even higher. How did a company that began over 60 years ago making wooden toys become such an international success?

The story begins with Ole Kirk Christiansen, a carpenter in the Danish village of Billund, who started a wooden toy company in the 1930s. Ole Kirk started out by making wooden cars, trucks, trains, airplanes, and animals. He continued to add toys to his line, and soon he was making wooden building blocks. He named his toy workshop Lego™, from a Danish phrase that means "playing well."

After World War II, when manufacturers began using plastics, Ole Kirk made a plastic baby rattle. In 1949, Lego™ produced its first plastic bricks. A toy trade magazine in Denmark warned that "plastic will never take the place of good, solid wooden toys." Fortunately Ole Kirk ignored its warning.

Ole's son Godfried Kirk Christiansen, known as GKC, was the one who developed the Lego™ system. GKC had an inspiration. He established criteria that he thought would mean certain success for toys he designed. One criterion was that the toy should be useable year-round. Another was that the toy should appeal to both boys and girls. One particularly clever criterion GKC thought of was that the more of the toy that people owned, the greater its value.

GKC searched through Lego™'s hundreds of products trying to find a toy that met his criteria. The toy he decided upon was a plastic interlocking brick. The company packaged an assortment of the little plastic bricks, and this new Lego™ system quickly became the company's best-selling product.

Not long after the company introduced the Lego™ system, it came out with a second line called Duplo™. Duplo™ bricks are eight times larger than the Lego™ bricks. This makes them easy for younger children to handle and too big for them to swallow. And, when children get older, they can use their Duplo™ bricks with Lego™ bricks because the two sets of bricks fit together.

In 1960 a fire destroyed Lego™'s wood toy factory, and after that the company concentrated on making only plastic bricks. Today the Lego™ company sells its bricks in more than 115 countries. The Danish town of Billund is now home to three Lego™ factories and Legoland Park, a theme park that features exhibits of Lego™ constructions. A 50-foot-high replica of Mount Rushmore stands in the park. It is made of 1.5 million Lego™ bricks and 40,000 Duplo™ bricks. A replica of the port of Copenhagen required 3 million bricks and two years of labor by eight model makers. Legoland is Denmark's biggest tourist attraction other than Copenhagen.

Evaulate

CHAPTER 8

Why Are There So Many Products That Do the Same Thing?

In the last chapter, you learned about the design process and technological problem solving. You learned that most designers use some sort of process as they design. They might not have the process recorded in an orderly fashion. Some might not completely be aware of their own design process. Yet most designers use a process that includes at least thinking, testing, and modifying.

If every designer uses some design process, how can there be so many different products on the market that serve the same purpose? This photograph of a parking lot illustrates just a small fraction of all the different kinds of cars in the world. Now think of how many different kinds of desks there are, or telephones, or watches.

In this chapter you will explore the relationship between design and diversity. You will study how criteria and constraints affect the diversity of products. Then you will be able to propose an answer to the question, Why are there so many products that do the same thing?

INVESTIGATION:
One Problem, Different Decisions

Think about the following question: How many ways can there be to make a cup of coffee, or to hold hair in a ponytail, or to attach two pieces of paper together? In this investigation you will analyze several different products that are designed to do the same thing.

Materials

For the entire class:
- the products your teacher provides

Procedure

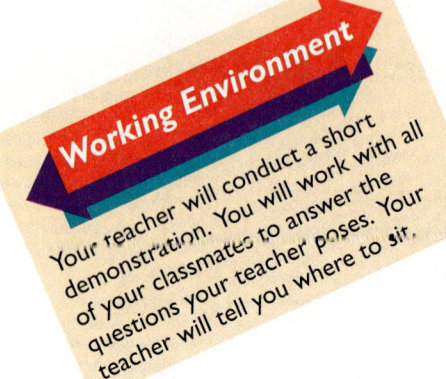

Working Environment
Your teacher will conduct a short demonstration. You will work with all of your classmates to answer the questions your teacher poses. Your teacher will tell you where to sit.

1. Observe the products your teacher displays.

   Notebook entry: Record your observations. Write descriptions of the products including the names of the items, their general purpose, and how they each accomplish their purpose.

2. With the rest of the class, discuss the questions your teacher asks you.

Wrap Up

Write an answer to the following in your notebook.

1. List two other products that have the same goal, but are different in the ways they look or the way they accomplish the goal.
2. Describe some of the different designs for the products you listed in question 1.

INVESTIGATION:
Shapely Designs

When architects and computer engineers design buildings or machines, they use basic building blocks, such as bricks or computer chips, respectively. Whether they design a building or a computer, the end product depends on the purpose or the function of the building or computer. It also depends on the availability and cost of materials. In this investigation you will design various things using different shapes as building blocks and different numbers of those shapes. You then will compare your designs with the designs from other groups.

Materials

For each team of three students:

- 1 copy of each of the following:
  Art-1
  Furniture-1
  Robot-1
  Art-2
  Furniture-2
  Robot-2
- Shapes, as many copies as you need
- 9 sheets of construction paper, any color
- glue stick or school glue
- 3 pairs of scissors

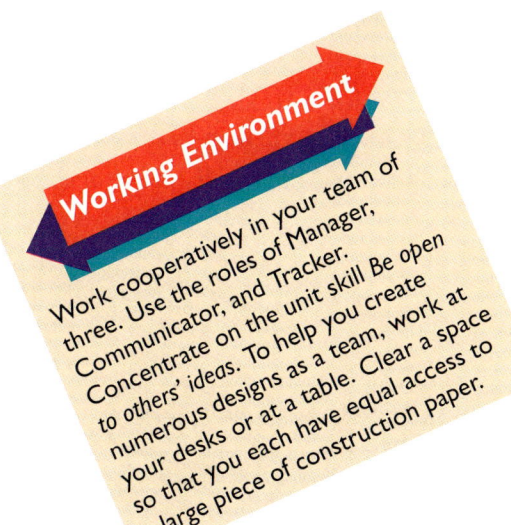

Working Environment

Work cooperatively in your team of three. Use the roles of Manager, Communicator, and Tracker. Concentrate on the unit skill. Be open to others' ideas. To help you create numerous designs as a team, work at your desks or at a table. Clear a space so that you each have equal access to a large piece of construction paper.

Procedure: Part A—Seven-Piece Designs

1. Obtain the materials.
2. Cut out the shapes from the Art-1 page.

   This page has 7 shapes.

3. As a team decide how to arrange your shapes to design a work of art.

   Your main objective is to arrange the shapes in a design that is pleasing to the eye. You must use all 7 shapes.

4. After you have agreed on an artistic design, glue the shapes in that design onto a piece of construction paper.

   Label the design with the Team Members' names, the date, and the word "Art."

5. Cut out the shapes from the Furniture-1 page.

   This page has 7 shapes.

6. As a team decide how to arrange the shapes to represent furniture in a family room or living room.

   You must use all 7 pieces.

7. After you have agreed on a design for a family or living room full of furniture, glue the shapes in that design onto another piece of construction paper.

   Label the design with the Team Members' names, the date, and the word "Furniture."

8. Cut out the shapes from the Robot-1 page.

   This page has 7 pieces.

9. As a team decide on a way to arrange the pieces to create a robot.

Explore — Why Are There So Many Products That Do the Same Thing?

You must use all 7 pieces. Remember to be open to the ideas of your teammates.

10. After you have agreed on a design for a robot, glue the shapes in that design onto another piece of construction paper.

    Label the design with the Team Members' names, the date, and the word "Robot."

11. Compare each design you made with 7 shapes with each design from the other teams.

    Notice the major similarities and differences among the designs.

12. Answer the following questions in your notebook.

    a. When you compared your designs with other teams' designs, which designs looked most like one another: the art designs, the furniture designs, or the robot designs?

    b. Which designs looked the most different from one another?

    c. Why did the procedures you used for each design make your designs look similar to or different from the other teams' designs?

Procedure: Part B—15-Piece Designs

1. Using the 15-piece design pages, Art-2, Robot-2, and Furniture-2, design a work of art, a family room or living room full of furniture, and a robot.

This time, however, you must use all 15 pieces (no more and no less) that we provide for each design. You cannot repeat any of the 7-piece designs previously created by any of the teams.

2. Glue your designs onto pieces of construction paper as before.

   Label each design with the Team Members' names, the date, and what the design represents.

3. Compare your designs to those of the other teams.

   Notice the similarities and differences among the designs from team to team.

4. Compare your team's 15-piece art designs to your team's 7-piece art designs.

   Notice the similarities and the differences.

5. Compare your team's 15-piece and 7-piece furniture designs.

6. Compare your team's 15-piece and 7-piece robot designs.

7. Discuss the following questions as a team.

   Notebook entry: Record your team's answers.

   a. In the 7-piece designs you created in Part A, the team's art, furniture, and robot designs probably showed different degrees of diversity. Did your 15-piece designs show the same pattern of diversity?

   b. From team to team, did the designs show more or less diversity

   ■ in the 15-piece art designs compared with the 7-piece art designs?

   ■ in the 15-piece furniture designs compared with the 7-piece furniture designs?

   ■ in the 15-piece robot designs compared with the 7-piece robot designs?

   c. How do you explain these patterns of diversity?

   d. Do you think that your own team's 15-piece designs were more or less creative than your 7-piece designs? Why might this be so?

   e. Do you think the members of your team are being open to one another's ideas? Explain your answer.

Procedure: Part C—Unlimited Designs

1. You again will design a work of art, a family room or living room full of furniture, and a robot. This time, however, use the shapes on the page titled Shapes.

   Use as many shapes for each design as you choose, but do not repeat any of the designs your team previously created in Parts A or B.

2. Glue your designs onto construction paper.

   Label them with the Team Members' names, the date, and what the design represents.

3. Compare your team's new designs to the new designs of other teams.

4. Compare your team's new designs to your team's 15-piece and 7-piece designs.

5. Discuss the following questions as a team.

   Notebook entry: Record your team's answers.

   a. Among the teams in the class, did the new designs follow the same pattern of similarity and diversity as the 15-piece or 7-piece versions? Why or why not?

   b. In general was there more diversity across teams with these new designs than there was with the 15-piece designs? Explain these results.

   c. Within your team do you think that your new designs were more or less creative than your 15-piece designs? Why do you think this was so?

Wrap Up

Write answers in your notebook to the following questions after you discuss them as a team. Prepare to share your answers and observations from all three parts of this investigation with the rest of the class.

1. Think of a reason that this investigation did not require you to make a work of art with 7 pieces, then a room full of furniture with 15 pieces, and then a robot with an unlimited number of pieces.

2. In your team of three, compose a paragraph that summarizes what you discovered in this investigation. You should include a discussion of these points:

   - How do more detailed criteria affect the similarity or diversity of designs different people produce?
   - How do constraints limit the similarity and diversity in those designs?

3. Prepare to share or explain your paragraph with the rest of the class.

4. On a scale of 1 to 10 (10 being the best) rate your team on how well you practiced the unit skill of being open to others' ideas. Take turns pointing out at least one instance in which each of you showed openness to a teammate's idea.

READING:
Similarity and Diversity in Designs

When you look at a cluster of buildings, a parking lot full of cars, a shelf full of books, or other groups of products, within the particular group, they might look very similar. There are times, however, when products within a particular group look very different.

Take athletic shoes, for example. If you went into a shop that specializes in basketball shoes, you would see different brands of shoes, but in general, they all would look basically the same and have the same basic features. If you went into a shop that specializes in a variety of athletic shoes, however, you might see a wide range of shoes that resemble each other much less. Low impact aerobic shoes look very different from basketball shoes, which don't look much like running shoes, which also do not resemble tennis shoes. Yet we still think of all of these shoes as part of a group of products that we call athletic shoes.

What is it about designers and the design process that accounts for the degree of diversity in groups of products? How much diversity you observe really hinges on the decisions that designers make during the design process. Because designers make many decisions as they develop a product, design is a creative process.

Stop and Discuss

1. What are some of the decisions a designer has to make when designing a product?
2. Complete the following statement by filling in the blanks using the words "more" or "less."

    The more decisions a designer is able to make, the _____ creative he or she can be with the design of the product, and the _____ diverse that product will be.

Think for a moment about the investigation you just finished—Shapely Designs. In that investigation, you used shapes to design a number of things. When you could design anything that looked pleasing to the eye, the teams probably developed a wide diversity of designs.

When you used the shapes to design a robot, however, the class designs were likely more similar to each other than the art designs were. Also your furniture designs were probably more similar to other teams' furniture designs. This pattern is apparent because the procedure directed you to use the shapes for a more specific purpose when you designed a robot or furniture than when you designed art. Art also has a purpose, but it has fewer limitations than a robot or furniture.

In general the more exact you are in defining the *function* of the product you design, the more similar that product will be to other products with the same function. Describing the function of the product is the same as listing the criteria for the product. So in the more exact you are about the *criteria* of the product you design, the more similar that product will be to other products with the same criteria.

For example, if you ask an automotive engineer to design a vehicle and do not list any specific criteria for the vehicle, you are likely to get a transportation machine of almost any type. However, if you ask the engineer to design a vehicle with these two criteria—it must transport six people and have enough cargo space for six suitcases—you probably will get something close to a station wagon or a mini-van.

One short statement makes this point clear: *Criteria limit the diversity of a design.* In the previous example involving car design, when the criteria became more specific, the engineer had fewer choices to make. In other words, the more criteria there are for a product's design, the less creative the designer can be.

Criteria, however, are not the only aspects of design that limit a designer's creativity. In Chapter 7, you learned that criteria *and constraints* affect the final product decisions a designer makes. Think back again to Shapely Designs. You probably noticed this general trend in your team's designs: the more pieces you were allowed to use for each design, the more creative you could be.

Recall from Chapter 7 that materials are a design constraint. Using only seven pieces constrained your designs much more than when you could use as many pieces as you wanted. The more pieces you had, the more creative you could be. The more constraints you have, then, the less creative you can be, and the less diverse your product will be. We could extend our previous phrase to say: *Criteria and constraints limit the diversity of a design.*

Making decisions involves creativity. The more choices you can make for your product, the more creative you will be in the design. When you have fewer choices to make for your product, your creativity is limited, and your product has a better chance of looking like someone else's product designed for the same purpose.

Stop and Discuss

3. Go back to the previous Stop and Discuss and check your answer to question 2. If you feel that you should change it, do so now. Explain the statement as you have completed it.

4. Think back to One Problem, Different Decisions. Working individually decide whether the products your teacher showed you could be any more diverse and still function the way they are supposed to. Then decide whether the designs were limited strongly by criteria and constraints or whether the designers were able to make many creative decisions. Justify your decisions and write your decisions and justifications in your notebook. Your teacher might call upon you to present your decisions and justifications to the class.

INVESTIGATION:
Up, Up, and Away!

If you have ever had a chance to go to an air show or visit an airport, you probably noticed that airplanes come in many shapes and sizes. Why do you think there is such a great diversity of airplanes? It has something to do with the idea you just read about: *Criteria and constraints limit the diversity of a design.*

In this investigation you will work in teams to design airplanes. You will have to adhere to specific criteria and constraints in doing so. When you are finished, you will compare your team's airplane design with those of other teams.

Materials—Part A

There are no materials.

Materials—Part B

For the entire class:
- 1 roll of masking tape
- 1 roll of transparent tape

For each team of three students:
- 10 sheets of 8½-by-11-in. paper, unlined
- paper clips
- 1 meter stick or metric measuring tape
- 3 pairs of goggles

Materials—Part C

For each team of three students:
- any materials from Part B
- 3 sheets of construction paper
- 1 bottle of school glue or 1 glue stick
- 10 rubber bands
- 3 craft sticks
- 10 sheets of 8½-by-11-in. paper, unlined
- any other materials your teacher provides
- 3 pairs of goggles

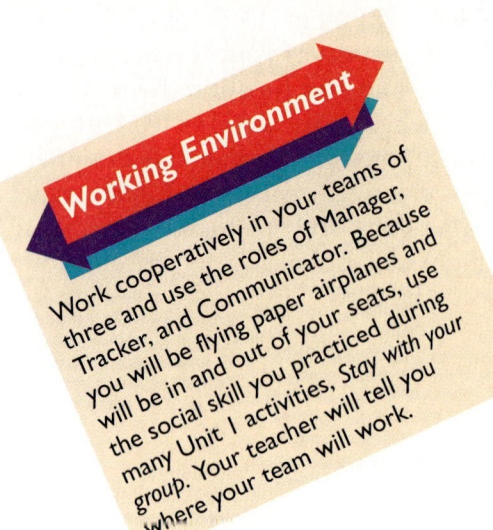

Working Environment

Work cooperatively in your teams of three and use the roles of Manager, Tracker, and Communicator. Because you will be flying paper airplanes and will be in and out of your seats, use the social skill you practiced during many Unit 1 activities, Stay with your group. Your teacher will tell you where your team will work.

Procedure: Part A—The Social Skill

1. Discuss the meaning of staying with your group as it pertains to this investigation.

   Remember always to read through the entire procedure first.

2. Turn back in your notebooks to the place where you created a T-chart for this skill with your Unit 1 team.

3. Share the ideas from all your T-charts.

Procedure: Part B—The Challenge

1. Read each of the following sets of design criteria and constraints.

   Design Plan 1

   Design a paper airplane according to the following criteria and constraints.

   Design Criteria: The airplane must carry a cargo, have wings, and travel at least three meters.

   Design Constraints: You must use one sheet of 8½-by-11-inch paper to make the airplane. Cargo must be represented by two paper clips. Using transparent tape is optional.

Design Plan 2

Design a paper airplane according to the following criteria and constraints.

Design Criteria: The airplane must have wings, travel at least four meters, and land as close as possible to a straight line of masking tape on the floor.

Design Constraints: You must use one sheet of $8\frac{1}{2}$-by-11-inch paper to make the airplane. Using transparent tape is optional.

Design Plan 3

Design a paper airplane according to the following criteria and constraints.

Design Criteria: The airplane must be as acrobatic as possible.

Design Constraints: You must use one sheet of $8\text{-}\frac{1}{2}$-by-11-inch paper to make the airplane. Using transparent tape is optional.

2. Predict which design will lead to the greatest class diversity and which will lead to the least class diversity.

   Think back to Shapely Designs and to the reading Similarity and Diversity in Designs. Justify your predictions and write your predictions and justifications in your notebook.

3. Share your predictions with the rest of the class.

4. Check your predictions by designing each type of paper airplane.
   Use the following steps to design your plane.

   a. Follow your flow chart of the design process from Chapter 7 as you design and build your plane. Revise your chart if necessary.

      Notebook entry: Record the decisions you make at each step.

   b. When you are satisfied that you have designed the best possible airplane for each design, put it aside in the area your teacher specifies.

      There will be a separate area for Design 1, Design 2, and Design 3. Be sure you put your names on your plane for the upcoming contest.

   c. Each person on the team should be able to explain the process that your team used to design your plane.

   ▲ **CAUTION: You must wear your goggles for step 5.**

5. Participate in each design contest according to your teacher's instructions. You will have a separate contest for each design.

Before you send your plane through the air, your teacher will select one person from your team to explain your team's design process.

Wrap Up: Part B

Complete the following tasks and answer the questions as a class.

1. Compare the planes from Design Plan 1 to the planes from Design Plan 2, the planes from Design Plan 2 to the planes from Design Plan 3, and the planes from Design Plan 1 to the planes from Design Plan 3. Which pair is the most different? Which pair is the most similar?
2. Where did you notice a greater diversity—among the planes of a certain design, or among the planes from different designs? How do you explain the diversity you observed?
3. Do your observations verify your earlier predictions?

Procedure: Part C—Free Designs

1. As a class choose the criteria from Design Plans 1, 2, or 3 to use for designing another plane.
2. This time, use only this constraint: The main body of the plane must be paper.

   You can use any kind or size of paper and any other materials that you wish. Every team in class should use the criteria from Design Plan 1, 2, or 3 according to the class decision.
3. Design your plane by following your flow chart for the design process.

   Notebook entry: Record all the decisions you make. Make changes to your flow chart if necessary.
4. Once you are satisfied that your plane meets the criteria that the class chose, put it aside for the flying contest.
5. Take turns presenting and flying your team's plane as your teacher directs.

> ▲ **CAUTION:** Wear your goggles and follow your teacher's instructions as you fly your planes.

Wrap Up: Part C

Answer the following questions as a class.

1. How much diversity did you observe among the planes you and your classmates just flew, each of which was designed with the same criteria? Explain your observations.
2. How would you rate your team's ability to stay together as a group: excellent, good, fair, poor?

3. List one strategy you will use in the next investigation to improve your rating for this skill.

CONNECTIONS:
Explaining Design Diversity

In the investigations Up, Up, and Away! and Anchors Away!, you gained much experience designing products. Now use this experience to convince someone who knows nothing about design that *criteria and constraints limit the diversity of a design*. Work individually and use any of the methods below to accomplish your task.

- Write a paragraph and read it to the class.
- Produce a radio or television commercial (you can ask other students to join you).
- Write a play and present it to the class (you can ask other students to join you).
- Create a pictorial diorama.
- Create an advertising brochure complete with photographs or sketches.

Materials

For the entire class
- assorted art supplies

 CHAPTER

Masters of Design

Consider yourselves designers! You now have reached the end of Unit 2, in which you focused on technological problem solving. Technological problem solving means using technology to help solve problems or to design products that extend human limits. In this chapter you will review what you have learned about technology and design. You also will have a chance to use your knowledge to evaluate something at your school and to redesign it for a different purpose. This will be your chance to change your school!

READING:
Let's Talk Technology—Again

Technology is a complicated sounding word, but by now you should be familiar with the topic. After all, you have designed boats, toys, works of art, furniture, robots, and airplanes. Now consider exactly what you have learned about technology.

First you learned that product designers must set goals for their product. They base these goals on what they feel would make their product the best at accomplishing a particular job. These goals are called criteria. After designers set their criteria, they must determine which factors might limit their ability to create the final product. These limiting factors are called constraints and usually have to do with types of materials, cost of materials, time, budget, and human factors. Once designers have identified their criteria and constraints, they make a series of final decisions that lead to the development of a final product.

Stop and Discuss

1. What criteria did you and your classmates use to evaluate and rank paper towels?
2. What criteria did you and your classmates use to evaluate and rank breakfast cereals?
3. List the constraints you think the designers of paper towels and the designers of breakfast cereals worked with when they designed their products.

In Chapter 7 you explored the process that you followed as you designed boats and toys. Your process allowed you to decide, build, test, and modify your product much in the same way designers do when they make a product.

Stop and Discuss

4. Review your flow chart of the design process. How did this chart help you design toys and paper airplanes?

Next in Chapter 8, you explored how criteria and constraints affect the creative process of design. You can account for the similarities and differences you see in many products by the number of criteria, constraints, and choices a designer was able to make. The more choices designers can make, the more diverse the products tend to be.

Stop and Discuss

5. Consider the criteria and constraints you used when you designed art, robots, and furniture in Shapely Designs. How did these criteria and constraints affect your designs?

In this chapter you will find out how comfortable you are with the topic of technology by completing the following two activities. These activities will test the limits of your design skills, so be sure to draw on all of your previous experiences and knowledge about technology.

CONNECTIONS:
Evaluating Your Environment

Everything around you—desks, chairs, lockers, drinking fountains, playground equipment, study areas in the library, and lunchroom equipment—was designed by someone. How well did the designers do? Did they account for the human factors of middle school students? In this connections section, work in your team of three to evaluate the design of something in your school environment. Follow these steps in conducting your evaluation. Then report your findings to the class.

1. In your team conduct a brainstorming session to choose a play, work, rest, or eating environment in your school to evaluate. Possible areas include the following: playground equipment, a study area in your library or classroom, the restroom, a portion of the lunchroom, a commons area in your school, an area of the gym, a locker room, or a waiting area in the office or health room.

2. Evaluate the environment that you've chosen. Your evaluation should answer the following questions. Your evaluation then can include whatever else your team feels is necessary.

   a. What criteria did the designers have in mind? (Although you cannot be certain about these criteria, you probably can make some good guesses.)

   b. What constraints were the designers working with? (Again, although you cannot know all of the constraints, such as the exact cost of materials or the budget, you can make some guesses about them. You should be able to figure out many of the constraints by pretending you had to design the environment.)

   c. How well did the designers accommodate human factors?

   d. How might you improve the designs?

Explore ■ *Explain* ■ *Elaborate*

e. If you were in charge of the school, would you buy similar equipment or look for something new? Why?

3. Prepare a report for the class. Include your answers to the questions in step 2. Each Team Member should write a portion of the report and take part in the presentation to the class.

INVESTIGATION:
Enabling the Disabled

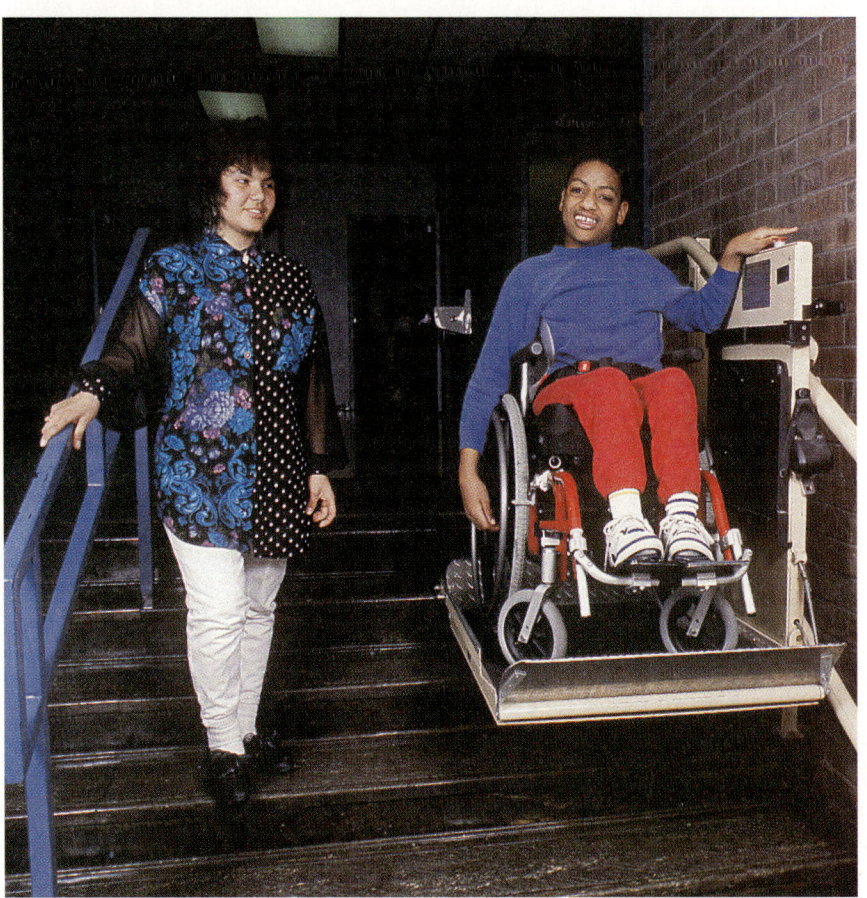

You probably have seen products that help handicapped people use environments that would otherwise be inaccessible to them. Close your eyes for a few moments as your teacher guides you through two situations in which you imagine that you are disabled, trying to cope in an environment that does not account for your needs. *Now stop and imagine . . .*

In this investigation you will redesign the environment you chose in Evaluating Your Environment to make it more accessible to handicapped students.

Working Environment

This is the last time you will work cooperatively with your current Team Members. Remember that all teammates should fulfill the duties of a Team Member. In addition, use the roles of Manager, Communicator, and Tracker. Be open to others' ideas and stay with your group. (Don't forget to implement the strategy you decided on in the Wrap Up to Up, Up and Away!) You first will meet in your threesome at your desks, and then your team will work in the environment you will be redesigning. Be sure during this time to stay with your group.

Materials

For each team of three students:
- 1 large sheet of poster board or large drawing paper
- 6 markers, assorted colors

Procedure

1. As a team conduct a brainstorming session to decide on a handicap that you will want to accommodate in the environment you chose in Evaluating Your Environment.

   The handicap could be one that someone at your school has, one that a friend or relative you know has, or a handicap that you are curious about.

2. Conduct your research of the handicap you choose.

   If you are or any Team Member is unsure about how to conduct research, read How To #4, How to Conduct a Research Project.

3. Return to the environment you evaluated in Evaluating Your Environment.

4. Make a list of the human factors that you now need to consider to accommodate a student with the handicap you researched.

   Notebook entry: Record your list.

5. Redesign the environment to accommodate the handicap you chose.

   Use the same criteria and constraints that you identified when you evaluated the environment before, and consider the human factors you identified in step 4 of this procedure. The Tracker should make sure that your team follows the flow chart for your design process.

6. Sketch your final design on a piece of drawing paper or poster board.

   Notebook entry: Use your notebook for any of your rough drafts. Take your time with your final sketch.

Wrap Up

Complete the following as a team.

1. Present your design to the class. In your presentation identify the changes you made in the environment and explain why you made these changes. As before all Team Members should participate in the presentation. After you finish ask the class for their suggestions about how you could modify your design to better accommodate the handicap you chose.

2. Discuss how well you think your team has worked together during this unit and how well your Team Members used the

unit skill of being open to others' ideas. Compose a page-long summary about how you know whether your team was or was not successful at cooperative learning and at using the unit skill. Be sure to include the following in your summary:

a. What can each of you do to continue practicing the Unit 2 skill in your new Unit 3 team?

b. How can you be successful in using the following social skills in your new Unit 3 team?

- Use your teammates' names.
- Speak softly so only your teammates can hear you.
- Stay with your group.

CONNECTIONS:
What Is Technology?

As an individual complete the following task and prepare to present the result to your class.

Pretend you are a writer at a dictionary company. Your job for the day is to write a clear definition of the word **technology**. You can check the dictionary in your classroom or library for a general idea of what kind of format to use, but do not use the dictionary definition. You must base your definition on what you learned in this unit. It also should include examples of concepts taken from everyday life. Write your definition in your notebook.

Materials

For the entire class:
- 30 dictionaries (optional)

SIDELIGHT

Putting Human Factors to Use for the Handicapped

You probably have seen a sign designating a parking space for the handicapped. The sign shows a profile of a human figure in a wheelchair. These signs are used to reserve close-in parking for handicapped drivers or riders. It's easy to assume that a car parked in a handicapped person's parking place is there because one of its riders is handicapped. But many times it is the driver of the car who is handicapped and often in a wheelchair. Did you ever wonder how a person in a wheelchair could get in a car and drive it? Well, thanks to modern technology, people in wheelchairs do drive. It all depends on how the car is designed.

One important criterion in the design of vehicles for people who use wheelchairs is "ease of access." The designers have to assume that the handicapped driver will be alone when he or she enters the car and that this driver will have to transfer his or her body from the wheelchair into the car and then store the wheelchair. The manufacturers of these vehicles decided to make cars that have two doors rather than four so that the doors can be extra long and to make the doors at least 36 inches high (the height of most wheelchairs). This gives the handicapped driver enough room to get in and then pull in the wheelchair.

One important constraint in designing a vehicle for the handicapped is the extra space required for the additional hand controls they need to drive the car, because often the handicapped driver cannot use his or her legs. Based on that constraint, manufacturers decided to make handicapped vehicles full-sized rather than compact-sized.

Other decisions that the manufacturers of these vehicles made based on the criteria and

This is what a steering column looks like with special equipment for the handicapped. Notice that the person does not need to use his or her legs to operate the accelerator or the brake.

constraints that account for handicapped people in wheelchairs include the following: an automatic transmission and power steering for easier handling, power brakes so that a person can apply them easily by hand, multi-adjusting power seats so the person can move the seats by pressing a button, cruise controls so the person does not need to press the accelerator arm all of the time, and power door locks and windows so the person does not have to lean over to operate the locks and windows of the car.

The manufacturers base all of these decisions on the associated human factors. The American Automobile Association publishes a 114-page book just for handicapped drivers. This book describes how drivers can find out where to purchase special equipment and learn about existing modifications in the designs of vehicles for the handicapped. Trying to accommodate people and their needs is what using human factors in technology is all about.

Engage ■ *Explore*

UNIT 3

Why Are Things Different?

You have completed two units about diversity. In the first unit, you explored how people were diverse in their limits. In the second unit, you learned how technology can help us overcome our limits. You spent quite a bit of time learning about designing things. As you designed products, you worked with many different materials. Materials usually constrained your designs. Why?

Why would it matter what materials you use to build a boat? Why can't all boats be made of wood? Even if all boats were made of wood, why do people use different varieties of wood to fulfill different criteria? Is there a difference in materials or even a difference in types of wood? Well, there must be, or considering materials wouldn't be such a big part of design.

In this unit you will explore materials in much the same way that professional scientists do. Scientists often want to know why materials are different. They study materials, conduct tests on the materials, make predictions about the materials, and conduct more tests on materials in an effort to explain what it is about materials that makes them unique.

This unit is about exploring the uniqueness of materials and trying to explain that uniqueness. As you develop explanations, you will learn how to make scientific explanations and how to test them. By the end of this unit, you should understand more about why things are different and more about using science to help you explain your observations.

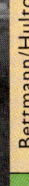
Bettmann/Hulton

COOPERATIVE LEARNING OVERVIEW

Marie is now used to working cooperatively in science class. She suddenly is faced with a situation outside of science class that relates to cooperative learning. After half a year of practicing social skills and learning to cooperate with others, you might be feeling more comfortable with the idea of working cooperatively on some science activities, too.

But what you have learned about working cooperatively isn't necessarily confined to your science class. You can try to apply the skills you have practiced to many other areas of your life. Take a moment now to discuss the characters' comments, then discuss the following:

1. Explain whether the situation that Marie is referring to would have been different if the students knew how to work cooperatively.
2. Explain how Al's comment applies to the situation.
3. In what other situations in your life could you apply the cooperative learning model and use social skills?
4. When does working cooperatively not apply?
5. In what future situations might it be beneficial to know how to work cooperatively?

You will be working cooperatively in a new team of three throughout this unit. Make sure that you and your Team Members take turns using each role. Also try to apply all of the social skills you have practiced in Units 1 and 2 as you work in your new team. For example, call your new teammates by name, try to be quick and quiet as you form your group, stay with your new team, and speak softly as you work. You will be using new skills for the activities as well as the new unit skill that Al mentioned: *Choose an explanation that includes the ideas of all Team Members.*

The first thing you should do with your new teammates is discuss this new unit skill and create a T-chart in your notebooks for it. With a T-chart you can always go back and modify or add to your ideas.

CHAPTER 10

Properties: The Material World

The title of this unit is Why Are Things Different? Before you can answer that question, you first will explore some of the diversity in materials. In Unit 2 you determined which properties of paper towels were important and then ranked different brands of paper towels according to those properties. This chapter will give you more practice in learning about some of the ways materials are diverse. You also will learn about some of the properties that make materials unique. You then will define and explore some common scientific properties of many materials.

INVESTIGATION:
Diversity of Bounceability

No, bounceability isn't a real word! But you probably can figure out what it means. It means how bouncy something is. What material is the bounciest stuff you know? Rubber? Plastic? The hamburger served in the cafeteria? What material is the least bouncy stuff you know? Wood? Clay? The hamburger served in the cafeteria?

In this investigation you will test different materials and rank them from most bouncy to least bouncy. But first you will observe a demonstration of bounceability.

Working Environment

Work cooperatively in your new team of three. Assemble your desks in a configuration that enables you to see each other. (You can use the same configurations that you did in your previous team of three.) Use the roles of Manager, Tracker, and Communicator. Practice the social skill Use your teammates' names.

Materials

For each student:
- 1 pair of goggles

For each team of three students:
- 1 wooden ball
- 1 glass marble
- 1 large rubber ball
- 1 steel ball (marble)
- 1 Styrofoam™ ball
- 1 wooden block (any size)
- 1 steel nail
- 1 wooden splint
- 1 rubber band
- 1 sheet of aluminum foil, 8-by-11 in.
- 1 sheet of clear plastic wrap, 8-by-11 in.
- various measuring tools of your choice such as a pan balance, a metric ruler, a piece of string

Procedure: Part A—The Social Skill

1. Write the first and last names of your teammates in your notebook.
2. Think of a strategy you will use at the end of this investigation to show the class how well you did in using your teammates' names.

   Here are some ideas: write a short rap using each others' names, introduce each other as if you each were famous, or make up a tongue twister or knock-knock joke for each of your names. You might have other ideas.

Procedure: Part B—Seeing Is Believing?

1. Turn your attention to where your teacher will conduct a demonstration.
2. Quietly observe the demonstration.

   Notebook entry: Record your observations.
3. Use your observations to participate in a class discussion about the demonstration.

Procedure: Part C—Bounceability Testing

1. Obtain the materials that your team will test for bounceability.

   You must test all of the items in the materials list for bounceability except the measuring tools. You may use any or none of the measuring tools as you test the items for bounceability.
2. Decide on an operational definition of bounceability.

   This means decide how your team will measure the bounceability of the materials. You might have to try a few things before you decide on a definition. It's okay to change your definition if it is not working. You will not need to compare or combine your results with those from other teams.

   Notebook entry: Record your operational definition.
3. Construct a data table for this investigation.

   As usual read through the entire procedure first before attempting to create a data table.

> ▲ **CAUTION:** Beyond this step and any time others around you are conducting bounceability tests, wear protective eyewear such as goggles.

4. Test the materials for bounceability.

   Use your operational definition and record a list of the variables that you control. Also record the results of your bounceability tests for each material in your data table.
5. Rank the materials from the most bouncy to the least bouncy.

   Use your results to determine this ranking.

   Notebook entry: Record your ranking.

Wrap Up

As a team compose a paragraph that summarizes this investigation. Each of you should write this paragraph in your notebook. You can include anything that you feel is important from this investigation, but be sure to address the following points:

- your operational definition,
- an evaluation of your operational definition and how useful it was in determining bounceability,
- the variables you controlled,
- an evaluation of your data table,
- the results of your tests, and
- how you ranked the materials for bounceability.

When you present your summary to the class, include your team's depiction of how well you used each others' names (Part A, step 2.)

READING:
How Things Are Different

You have been studying the diversity of things since the beginning of the year. First you explored the ways you and your classmates are diverse. Then you expanded your exploration to include the diversity of products. Now you will have a chance to study some of the properties of materials in order to answer the unit question, Why are things different?

Your first experience in studying a property was in the previous investigation, Diversity of Bounceability. The property you explored was bounceability or how "bouncy" certain materials are.

Stop and Discuss

1. What results in your investigation surprised you?
2. Explain whether or not items made of the same materials had the same bounceability.

As you might have discovered in Diversity of Bounceability, exploring properties sometimes can be a complicated process. A property might be influenced by many different factors. In the case of bounceability, the shape of an object influences the amount of bounceability almost as much as the material composing the object. Historically scientists have spent a lot of time identifying and defining specific properties of materials. One reason scientists have invested so much time in studying properties is because scientists are curious about the hows and whys of things. Another reason is that an understanding of properties helps designers construct better products. In fact technology might not have come this far if scientists hadn't studied the properties of things. Consider the following:

> In 1938 Dr. Roy Plunkett was studying the types of gases used in refrigerators and air conditioners in Du Pont's New Jersey laboratory. One evening he left his experiment, including containers filled with gases, out overnight. The next morning he found that one of the containers of gases was coated inside with a thin, solid material. He immediately began studying the material to identify its properties. He applied chemicals of all types to see whether any of them would cause the strange material to corrode or disintegrate. None of the chemicals did. He did other tests to determine the basic texture of the material and found that it was extremely slippery. He named this material for its chemical name, *tetrafluoroethylene*.
>
> Du Pont laboratories thought that it was an amazing material and continued to test it for other properties. Eventually they discovered another important property: the material retained its slipperiness even when it was heated. Du Pont decided that this material would be very useful in industry to coat certain tools and pieces of machinery.
>
> *The Guinness Book of World Records* listed the material as the slipperiest substance on earth, comparable to "the slipperiness between two ice cubes rubbing against each other in a warm room." The president of Du Pont in the 1950s, Marc Gregoire, decided to have his fishing tackle coated with the material to reduce the incidence of things sticking and tangling on it. Mrs. Gregoire was so impressed by the slippery property of the material that she asked her husband if she could have her pots and pans coated with the material. Maybe then food wouldn't stick so much during cooking, and washing dishes would be easier. Well, Mrs. Gregoire's idea worked. Today pans coated with the strange material are common in many households. If Dr. Plunkett had not decided to examine the strange material and identify its properties, today we might be without Teflon™.

Over the years, scientists have identified so many properties of materials that it would take numerous pages just to list them.

Stop and Discuss

3. List as many properties of materials as you can think of.

To get an idea of the types of material properties scientists have identified and how these properties differ from one another, consider three common properties: **translucence**, **hardness**, and **viscosity**. By yourself study Figures 10.1, 10.2, and 10.3 and read the property descriptions below each one. Record answers to the

Figure 10.1

Will three sheets of paper be more or less translucent than two sheets?

stop and discuss questions in your notebook. Your teacher will call on any class member to answer the questions in a class discussion.

Translucence

Isaac is right. Scientists measure the property of translucence by how much light passes through a specific material. The more translucent a material is, the better you can see the light passing through the material. For example, glass can have varying degrees of translucence. Perhaps you have seen bathroom windows or shower doors that are not very translucent. The glass that makes up the windshield of a car, on the other hand, is very translucent. Even cloth can be translucent. Some sheer curtains might let a lot of light in the house. Other curtains are made of cloth that hardly lets any light through. We say that sheer curtains are translucent, and that nonsheer curtains are less translucent or nontranslucent.

Stop and Discuss

4. Some materials are translucent only in certain forms. For example, when is wood translucent and when is it not translucent?
5. Identify a product in your home that is made of a translucent material.
6. Think of one reason why the designer of the product you listed in question 5 might have wanted the product to have the property of translucence.

Hardness

Figure 10.2

How would you define the property of hardness?

Hardness can be a measure of how firm something is. This means that hard materials resist pressure much better than soft materials do. That is why Ros and Issac are complaining about the marshmallows. They can't squeeze them, which means that they are resisting pressure. On the other hand, how would you compare the hardness of an emerald and a diamond, if both resist being squeezed? Scientists consider a diamond harder than any other material, because a diamond can scratch any material except itself and no other material, can scratch a diamond.

Stop and Discuss

7. Explain which method you could use to measure the hardness of marshmallows.
8. Identify a product in your home that is made of a hard material.
9. Think of one reason why the designer of the product you listed in question 8 might have wanted the product to be made of a hard material.

Explain

Viscosity

Marie has just supplied you with a good description of viscosity. How fast or how slowly a liquid pours indicates how viscous a material is. Often we think of viscosity as a measure of the thickness of a substance. This is partly true, but we would be more correct to include a measure of the liquid's "pourability" or "flowability" to accurately define its viscosity. A viscous liquid, such as Marie's paint, does not pour or flow as easily as a nonviscous liquid, such as Al's paint.

Stop and Discuss

10. Name the most viscous liquid you can think of.
11. Name the least viscous liquid you can think of.
12. Identify a product in your home that is made of a viscous material.
13. Think of one reason why the designer of the product you listed in question 12 might have wanted it to be viscous.

Now that you have basic definitions for these three properties, you will have a chance to work with different materials to explore these properties more fully. Refer to this reading whenever you are unsure about how to investigate one of these properties.

SIDELIGHT

Does Your Polyvinyl Acetate Lose Its Flavor on the Bedpost Overnight?

This is a portrait of Antonio Lopez de Santa Anna, the famous Mexican General of the Mexican Revolution: He sat for this portrait in 1858.

The Bettmann Archive

Have you ever heard of Santa Anna, the famous Mexican general of the 1800s? If you haven't heard of him, how could you find out who he was?

Well, in addition to being an important historical figure, did you know that Santa Anna is responsible for one of America's favorite vices? He used to like to chew chunks of chicle, which are pieces of sap from sopadilla trees that grow in the Mexican jungle. Santa Anna discovered that chicle has the amazing property of being chewy for a long time.

Thomas Adams, an American inventor to whom Santa Anna introduced chicle, decided to market it in small bite-sized balls. People were already chewing wax for pleasure, but they found that the rubbery property of chicle was far more satisfying, Adams's chicle balls became a success. Later, other inventors thought of adding flavor to the chicle. Can you imagine chewing gum that has no flavor? Adams followed suit and introduced chicle that had been flavored with essence of licorice. He called his licorice-flavored chicle Black Jack™ gum. Black Jack™ gum, dating back to the 1870s, is still manufactured and enjoyed today and is the oldest brand of gum on the market. The rest is history. The most successful person to jump on the bandwagon was William Wrigley, Jr.

Modern supermarket shelves contain a bewildering variety of gums. The gum you chew today is not Santa Anna's chicle, though. Because scientists study the properties of things, they have discovered an even better chewing material: polyvinyl acetate. This is the marvelous material that you chew, snap, pop, blow, and have to throw out during class time. Surprisingly no gum manufacturer has come out with a gum named Alamo. (If you don't get this joke, then you haven't found out who Santa Anna was.)

Explain

Properties: The Material World ■ **195**

INVESTIGATION:
Properties for Sale or Rent

In this investigation think of yourself as Dr. Plunkett, Mrs. Gregoire, or Tom Adams. *You* now have a chance to be a properties scientist. In the following investigation, you will study one of the properties that we described in the previous reading. You will have a number of materials to choose from to help you study this property. You and your peer scientists will decide how to test your given property, on what materials you will conduct your tests, and finally, how you will rank your materials.

Materials

For each team of three students:
- any materials you choose to test for your assigned property

Procedure: Part A—The Social Skill

1. Pay attention to the class discussion your teacher will conduct now.

   Be prepared to participate.

2. Answer the questions your teacher will ask you.

   Your teacher will tell you for which questions you should record answers.

   Notebook entry: Record the answers

3. Discuss your answers and observations with your teammates.

Procedure: Part B—Exploring Properties

1. In your notebook record the property your team is responsible for investigating.

   Follow the teacher's directions to determine which one property (translucence, hardness, or viscosity) your team will investigate.

2. Obtain the materials that you will test and rank.

3. Decide on an operational definition for how you will measure your assigned property in the materials you chose.

   The Communicator will need to check with other teams, because you will be combining and comparing data with the other teams that are investigating the same property. Remember that to combine and compare data you need to use a common operational definition. If you

Working Environment: Work cooperatively in your team of three. Use the roles of Manager, Communicator, and Tracker. Move your desks together in a threesome configuration, or work at a table. As you work practice the new social skill. Let others finish without interrupting them.

have trouble deciding on an operational definition, review the reading for ideas.

4. Construct a data table.

   This data table should include room for information about what your test is and how each material responds to your tests.

5. Conduct your tests.

   Remember that modifying your operational definition is perfectly acceptable and scientific.

   Notebook entry: Record the results of your tests in your data table.

6. Rank the materials you tested from the hardest to the softest, most viscous to least viscous, or most transparent to least transparent.

   Use the results of your test to do this. Do not compare your results with other teams' results as you rank materials.

Wrap Up

As a team complete the following according to your teacher's directions.

Combine your data with the data from other teams that investigated the same property and think of some way to represent your data in a graph or picture form. You may choose a traditional method such as making a line or bar graph, or you may think of a less traditional way to show the combined rankings in picture form. You will have successfully completed this task as long as you represent your data in some sort of picture and your picture explains the following points:

- the property you investigated,
- the ranking of materials for your property,
- your operational definition, and
- the variables you controlled.

Discuss the following questions with your classmates. In your discussion show how well your team uses the skill of letting others finish without interrupting them.

1. Think of the property you investigated. Of all the materials in the world, which material is the best example of that property?
2. Which material is the worst example?
3. When you compared data with the teams that investigated the same property you did, were the rankings identical? Explain your answer.

CONNECTIONS:
Is the Most Always the Best?

People have a tendency to think that "most" and "best" are synonyms; that is, that they mean the same thing. For instance, think about the rubber balls from the beginning of this chapter. One bounced very high, and the other bounced hardly at all. You might think that the one that bounced the highest was made out of the best materials. This is *partly* true—if the designer's idea was to make a bouncy object.

But what if the idea was to make an object out of rubber that had as little bounce as possible? Consider automobile tires. Designers have worked to develop tires that make an automobile ride as smooth as possible. Tire manufacturers use a special material called *poly(styrene-butadiene) co-polymer* to make tires because of its remarkable property of not bouncing on bumps in the road but absorbing the impact of the bumps instead. Because of that absorbing property, other manufacturers use the same material to line the large containers in which bomb squads store bombs. Why? If a bomb were to explode in storage, it would be in a container that would absorb a lot of the "force" of the explosion.

Can you guess what kind of rubber the little ball that didn't bounce was made of? You guessed it: *poly(styrene-butadiene) co-polymer*!

There are other examples of situations in which a material that has the "most" of a property is not the "best" choice for the job.

Write answers to the following in your notebook. Be prepared to share your answers in a class discussion. Think about your experiences with the properties of viscosity, translucence, and hardness.

1. Describe a situation when the most viscous food (perhaps ketchup or molasses) would not be the best choice.
2. Describe a situation when the most translucent material would not be the best choice.
3. Describe a situation where the hardest material is the best choice.

Consider the property of stickiness. There are times when "the most" is "best," and times when "the least" is "best." Give an example of when it might be better to use a material that was very sticky and an example of when it might be better to use a material that was only slightly sticky.

Evaluate

CHAPTER 11

Scientific Explanations Are Ancient History

Now that you have spent some time investigating *how* materials are different, it's time for you to start trying to find out *why* materials are different. How can you figure out why some rubber is bouncy and other rubber is not bouncy? How can you figure out why some liquids are highly viscous and other liquids are not very viscous at all? For thousands of years people like you have been *trying to figure out why materials have different properties*. This chapter will help you begin to answer your *why* questions and to learn about how people long ago answered similar questions. It all begins with the following investigation.

INVESTIGATION:
Let There Be Colored Light!

In this investigation you will try to determine the contents of a light box without touching, shaking, or looking in the box. Using your keen powers of observation, you will develop your own explanation for what is inside the light box.

Working Environment
You will work as a class during this investigation. As others offer their opinions, try to be open to their ideas.

Materials

For the entire class:
- 1 light box
- 1 flashlight
- 2 sheets of unlined white paper

Procedure

1. Observe the light box demonstration.

   Your teacher will tell you where to sit for this step.

2. Describe what you observed.

   Notebook entry: Record your observations or sketch what you observed.

3. Prepare to share your own explanation of what is inside the light box.

   You can modify your explanation as you listen to the ideas of others.

Wrap Up

When your teacher has put away the light box, think about the following question and write an answer in your notebook.

- Do materials have insides the way boxes do?

INVESTIGATION:
Strange Phenomena

One way of answering the *why* questions is a lot like the way you tried to figure out what was inside the light box your teacher showed you. For the light box you might ask, "Why does one color of light come out of one hole and a different color come out of another hole?" You can answer that question if you can answer another question: what is going on inside the light box that makes us see what we see?

Think of the rubber balls again. The question we want to answer is why is one ball bouncy and the other ball not bouncy? We can answer that question if we can answer Marie's question in Figure 11.1.

Figure 11.1

What does Marie mean about the insides of rubber balls? Like Ros said, cut them open and there's rubber inside. Is this what Marie is referring to?

In order to find out whether or not materials have insides and even what those insides might be, you need to do some more exploring. In this investigation you will observe three different phenomena or "happenings." Your careful observations will be important to your understanding of the reading that follows this investigation.

Materials

For each team of three students:
Part B
- 3 medicine droppers
- 1 small beaker of water
- 1 small beaker of Solution A
- 1 small beaker of Solution B
- 1 lid or 1 base of a petri dish
- 1 sheet of black construction paper, 8½-by-11 in.
- 3 pairs of goggles

Working Environment

Work cooperatively in your teams of three. Create a work space using your desks or a table that allows each of you equal access to the materials as you use them. Use the roles of Manager, Tracker, and Communicator. Practice the social skill Let others finish without interrupting them.

Explore

Part C
- 1 tablespoon
- 1 plastic resealable bag
- 1 tbsp of mystery mix
- 1 tbsp of acetic acid (vinegar)
- 3 pairs of goggles

Procedure: Part A—You Can't Touch This

1. Pay attention to the demonstration your teacher will conduct.
2. Record your observations.

   Your teacher will ask you specific questions about your observations, which you should answer in your notebook.

Procedure: Part B—Puddle Mystery

1. Obtain all of the Part B materials.
2. Throughout this procedure count the number of times one person interrupts another.

   The Tracker should record the count in his or her notebook for this part and Part C.

> ▲ **CAUTION:** After this step you will need to wear eye protection.

3. Place the lid or the bottom of a petri dish on black construction paper.

   The Manager should do this.

4. With a medicine dropper, slowly drop water into the center of the dish until there is a puddle the size of a quarter.

   The Manager should do this.

5. Carefully fill a second medicine dropper with Solution A.

   The Tracker should do this.

6. Carefully fill the third medicine dropper with Solution B.

   The Communicator should do this.

> ▲ **CAUTION:** Avoid getting any solution on your skin or clothes. If you do, wash the area with cool water immediately. Then notify the teacher of any spills.

7. Hold the filled droppers opposite each other on the outer edges of the puddle, as shown in Figure 11.2.

   The Tracker and Communicator should do this. Be careful not to touch the puddle with the tip of the dropper.

Figure 11.2

Hold the droppers above the puddle on opposite edges. Remember not to touch the droppers to the puddle.

8. At the word "Go," squeeze four drops of liquid from each dropper onto opposite edges of the water puddle.

   The Manager will say "Go," and the Communicator and the Tracker will squeeze the droppers at the same time.

9. Observe what happens in the puddle of water.

   Notebook entry: Record your observations.

10. Return the materials for Part B, except for the safety goggles.

Procedure: Part C—Bubble, Bubble, Toil, and Trouble

1. Obtain the materials for Part C.

   ▲ **CAUTION:** After this step you will need to wear eye protection.

2. Seal a plastic bag halfway across the top.

   STOP: Is the Tracker keeping track of each time one person interrupts another?

3. Through the unsealed half of the opening, place one tablespoon of mystery mix into the bag.

4. Add one tablespoon of vinegar to the mystery mix and *immediately* seal the bag completely.

   Avoid getting any of these materials on your skin or clothes. If you do, rinse the area with cool water immediately.

5. Mix the materials by gently squeezing the bag.
6. Observe what is happening inside the bag and to the bag.
   Feel the bag with your hands as the experiment progresses.
7. Return the Part C materials.
   Dispose of the materials as directed by your teacher.

Wrap Up

Complete the following tasks as a team. When you have discussed your answers, record them in your notebook. Be prepared to share your answers in a class discussion.

1. For each phenomenon in Strange Phenomena you recorded some observations. Discuss the observations you recorded and any other observations your teammates can remember. Then try to explain what you think happened inside the materials you used. After each explanation you record, leave space in your notebook because you will modify your explanation later.

2. Have the Tracker anonymously add the number of your team's interruptions to the class count. Discuss how the class did in using the social skill.

READING:
The Chinese and Greeks Tell What Happened

Now that you've had a chance to explain what you observed, you might wonder how other people might explain why matches burn, why two clear liquid substances can combine to form white particles, or why a mystery mix and vinegar together produce bubbles and heat. For a long time, people have noticed that different materials have different properties, and for a long time they have been asking *why*.

Almost 3,000 years ago (1000 B.C.), Chinese philosophers tried to answer *why* questions. They wanted to explain why different materials, such as wood, water, and sand had different characteristics. Why did things smell different from one another? Why did wood burn easily, but sand did not?

It seemed possible to these philosophers that something might be going on inside the materials that could explain the things that people observe directly. It is a lot like the light box your teacher demonstrated. If you knew what went on inside the light box, you could explain why you saw what you did on the outside of the light box.

If you open the light box and look in, you can see what is *really* inside. But here's the catch: you can't open up a piece of paper, a

Figure 11.3

This is an artist's rendition of an ancient Chinese philosopher.

piece of wood, a piece of rubber, or a piece of any other material to see what is *really* inside. Sure, you can cut these materials apart to see more of the same material, but you can't see the inside of the material by cutting it open. This story is about figuring out what is inside things without ever being able to see what is *really* there. Chinese philosophers were among the first people to do this.

The Chinese philosophers decided that everything in the world was made up of just five things. They called these things "elements." The five elements were fire, earth, metal, water, and wood. The Chinese philosophers figured out a way to explain natural occurrences by observing changes in these five elements. For example, they would say that wood changes into fire and fire then changes into earth (ashes). Earth, they decided, then produces metal. Their evidence for this was that metals such as gold and iron are found in the ground. After observing dew clinging to metallic objects, they figured that metal must produce water. Finally, if you water an acorn in the ground, you eventually get a tree, so they reasoned that water produces wood. If these Chinese philosophers performed the experiments you just did, they would have used these five elements to explain the phenomena you observed.

About 500 years later (about 500 B.C.), Greek philosophers began developing their own explanations. Unaware of what the

Chinese were thinking about, they too began to wonder why things had different properties. A Greek philosopher named Thales (THAY leez), who lived between 634 and 546 B.C., thought that what was "inside" all materials was water. He thought this because water can take on many shapes—it can be ice, snow, liquid, or it can form clouds. Another Greek philosopher named Heraclitus (hair uh KLEE tus), who lived between 535 and 475 B.C., thought that fire was the basic substance that made up the world. Why? Because everything in the world was always changing, and to the Greeks, fire represented "that which moves."

A Greek named Empedocles (em PED oh cleez), who lived between 493 and 433 B.C. decided there were four basic things that made up all materials: fire, water, earth, and air. This idea caught on. Using these four elements, the Greeks could explain almost anything. They were able to develop explanations by reasoning that every material was composed of these four things, but not in the same amounts. For example, if they wanted to explain the properties of a piece of cloth, they might have said that it was made of the tiniest amount of fire, and that's why clothing keeps people warm. That same piece of cloth had a little bit more air in it than fire, which gives cloth its characteristic lightweight property. The movement of cloth, the way it clings and flows as it moves, might have suggested that a greater amount of water was in the cloth than air. Finally they might have said that the major element was earth, which gave cloth its solidness. They might have said that cloth was about 10 percent fire, 15 percent air, 25 percent water, and 50 percent earth.

Later the Greek philosopher Aristotle, (air ih STAH tul) who lived between 384 and 322 B.C., added to Empedocles's explanation

Figure 11.4

Many of the famous ancient Greeks were depicted with sculptures, like this bust of Aristotle.

by saying that each of the four materials, fire, water, earth, and air, was a combination of heat, cold, moisture, and dryness. In Aristotle's explanation fire was hot and dry, water was cold and wet, earth was cold and dry, and air was hot and moist.

The Greeks used Empedocles's and Aristotle's ideas over and over in almost any way they wanted to. Their explanations seemed to explain things for them. For about 2,000 years, most people believed that almost everything in the world was composed of fire, earth, water, and air.

CONNECTIONS:
Thinking Like the Ancients

Now imagine that you can go back in time, and suppose that you are either an ancient Chinese or ancient Greek philosopher who has just observed the phenomena you observed in the investigation Strange Phenomena. Explain these three phenomena using either the ancient Chinese or ancient Greek ideas. Use words you think the early philosophers might have used.

1. Part A—You Can't Touch This
2. Part B—Puddle Mystery
3. Part C—Bubble, Bubble, Toil, and Trouble

CHAPTER
12

Using Scientific Models to Answer Questions

What could be inside the presents in the chapter opening photograph? Suppose your birthday were coming up soon. Because you shouldn't open them until your birthday, you just look at your presents every day. Perhaps you look at the boxes and try to guess by their appearance what might be inside. How else might you determine what is inside each present without opening it?

Chances are you wouldn't learn much just by looking at your presents. Instead, you might conduct tests using your other senses. You could feel the shape through the wrapping paper, jiggle or lift the presents to see how heavy they are, or even smell the presents to try to obtain some clues. Then you might be able to guess what is inside. Yet how would you know whether your guesses were correct?

The strategies that you might use for guessing what a wrapped present contains are similar to the strategies scientists use to try and determine what is inside materials. In this chapter you will use scientific strategies to try to determine the insides of things. You also will learn how scientists use scientific explanations to answer their questions. Then you'll know whether scientists ever can be certain about their answers.

INVESTIGATION:
Mystery Box

In this investigation you will continue answering the question, Why are things different? You will do this by trying to figure out what is underneath a piece of cardboard. You won't be able to see or touch whatever is there. The process you use will help you understand how today's scientists developed their explanations for why things are different.

Materials

For each team of three students:
- 1 mystery box
- 1 strong magnet
- 1 sheet of light-colored construction paper
- 2 large marbles
- 2 small marbles
- 1 wooden ruler
- other materials you choose to use as needed

Procedure

1. Read the following rules for investigating your mystery box (see Figure 12.1).

   Make sure that each Team Member understands these mystery box rules. It is the Tracker's job to make sure that the team follows these rules.

2. Obtain your team's materials.

3. Without violating any of the rules, take a 5-minute turn using the materials to try to determine what is under the cardboard.

   The Communicator should take the first turn.

4. Observe the actions of the person investigating the box and what happens inside the box.

   Notebook entry: Record your observations.

5. At the end of 5 minutes, stop trying to determine what is under the cardboard.

   The Tracker is the timer except when he or she is busy with other duties. At that point he or she should assign the job of timing to another Team Member.

6. Draw in your notebook what you think is under the cardboard.

   All Team Members should do this. It's okay to guess!

7. Take turns showing and describing what you think is under the cardboard.

   This includes making a sketch of what you think is under the cardboard after each person investigates the mystery box.

Working Environment

Work cooperatively in your team of three. You will need a work space that allows each of you equal access to a large box. Use the roles of Tracker, Manager, and Communicator. Concentrate only on the unit skill Choose an explanation that includes the ideas of all Team Members.

Figure 12.1

Refer to these rules as you try to determine the contents of your mystery box.

> **RULES**
>
> To try to determine what is underneath the cardboard for the investigation Mystery Box, you MAY do any of the following things:
>
> 1. Shoot marbles or objects at whatever is under the cardboard using a ruler, a pencil, or a pen.
> 2. Bring the magnet close to the cardboard inside or outside the box or gently touch the magnet on the cardboard.
> 3. Move the box around and lift the box off the table or desk in order to move the marbles around.
> 4. Do anything else that does not violate the rules that follow.
>
> You MAY NOT do any of the following things:
>
> 1. Touch what is under the cardboard, either directly with your fingers or indirectly using objects other than marbles.
> 2. Place anything, including the magnet, a ruler, or your fingers under the cardboard. (If a marble gets stuck, ask your teacher to dislodge it.)
> 3. Lift, push, pull on, or grab the cardboard in any way.
> 4. Poke holes in the box that will allow you to see under the cardboard.
> 5. Peek through any existing holes in the box to see what is under the cardboard.
> 6. Hit the cardboard with anything. This includes knocking on it and dropping marbles on it.

8. Repeat steps 3 through 7 until each Team Member has had a 5-minute turn to use the materials.

   During each turn all Team Members should record their observations and draw and describe what they think is under the cardboard.

9. Continue taking 5-minute turns investigating the mystery box then drawing and describing what you think is underneath the cardboard until your teacher tells you to stop.

Wrap Up

Complete the following tasks and answer the questions before your class discussion.

1. As a team decide on one description for what is under the cardboard. Draw a picture of your final decision on a sheet of construction paper. Make sure everyone on your team can explain to the class what your team thinks is under the cardboard and why. That is, be ready to explain the evidence you have that your team's drawing is correct.

2. Present your team's drawing of what you think is under the cardboard in the mystery box and explain your idea by telling the class what evidence you have that supports your idea. Show the class how well you used the unit skill by pointing out where you included the ideas of all teammates in your explanation.

3. Answer the following questions in a class discussion:
   a. Did each member of your team have the same evidence with which to determine what was under the cardboard?
   b. Did the members of your team agree each time on what was under the cardboard? Why or why not?
   c. Did all of the different teams agree on what was under the cardboard? If not, explain why not.
   d. What reasons might explain why different teams agreed on what was under the cardboard?
   e. Suppose you could never lift up the cardboard and discover what was really in the mystery box. Could you prove to a friend, without a doubt, that what you thought was under the cardboard was really there? Explain your answer.

READING:
Another Explanation

How Explanations Change

You already have learned how some of the ancient Greeks and Chinese explained what made things different. They decided that a few basic elements such as earth and fire made up the universe. Other philosophers and scientists throughout history developed different explanations for what was inside of things. One idea that people keep coming back to was the notion that if you break materials up into smaller and smaller pieces, you will eventually find particles that you can't break apart. One of the first philosophers to develop an explanation like this was Anaxagoras (an ax uh GOR us), who lived around 450 B.C. His explanation went something like this:

> All materials are made up of an infinite number of very small "seeds." These seeds are mixed together in a material like a mixture of different colored sand. The properties a material has (such as hardness, color, and smell) depend on how many of each kind of seed are in the material.

A philosopher named Democritus (de MOK rih tus), who lived about 100 years after Anaxagoras, had a different explanation:

> All materials are made up of very small objects called "atoms." (The Greek word atom means "indivisible.") These atoms are the smallest things that exist. Different materials are made of different kinds of atoms. Some atoms are rough, others are smooth. Some are large and others are small. Atoms don't have color, taste, or smell by themselves. The motion and arrangement of the atoms in a material gives a material a certain color, taste, or smell.

Scientists' current explanation for what is inside of materials is based on Democritus's explanation of the atom, but it's more sophisticated. Some of the people throughout history who elaborated on Democritus's ideas include Isaac Newton, John Dalton, J.J. Thompson, E.J. Rutherford, Marie Curie (her picture is in the unit opening collage of photographs), and Niels Bohr. Perhaps you've heard some of the terms people use when talking about atoms, such as protons, neutrons, electrons, nucleus, neutrinos, or gamma rays. People use different names to describe different kinds of particles, such as atoms, ions, and molecules. You might study types of particles in the future. For now just focus on the idea that materials are made up of particles.

Two of the first questions you might ask about particles are: how do we know that materials are made up of particles and can we see them? Yes, it is possible to observe larger particles with an electron microscope, which provides evidence that materials are made of particles. We cannot see smaller particles, though, which makes it difficult to gather evidence about them. Thinking about the mystery box investigation might help you realize how scientists gather evidence about things they cannot see.

Stop and Discuss

Consider the process your team used to try to determine what was under the cardboard. We will now refer to what was under the cardboard as the target. One of the first things you might have done was to pick up the box, shake it, and listen to the sounds it makes.

1. Review your first idea about what the target in the mystery box looked like.

   You then used the materials to determine whether or not your first idea was correct.

2. If your first idea was correct, what *should* have happened when you fired a marble at the target? Sketch or describe what your prediction would have been.

3. Sketch or describe what actually happened when you fired a marble at the target.

4. Explain how this new evidence changed what you thought the target looked like. Sketch or describe how your idea changed.

5. How many times did your idea about what the target looked like change? (Think about each time a different person conducted tests with the materials and whether or not he or she found new evidence that changed your mind.)

This type of process for determining what the target looked like could have gone on and on. You could have decided what the target might be, and then *tested* your idea. Each time you tested your idea, you might have changed your mind about what the target looked like. You then tested your *new* idea to verify it, but you might have discovered new evidence that again changed your mind. Some teams might have changed their ideas over and over again until the teacher called a halt to your investigation of the mystery box.

The teams could have continued testing their ideas until they were quite certain what the target looked like without ever seeing it. They then might have concluded something like this: The results of all our tests indicate that the target has a certain shape and is made of certain materials, so we are confident that the target really has that shape and those materials.

Firing marbles at an unseen target is similar to experiments that scientists conduct today to learn about particles. They fire small particles at materials to see what happens. What happens to those particles is consistent with the idea that the material itself is made of small particles. Some of the fired particles pass through the material, which means there might be empty space in the material. Some of the fired particles bounce off at angles (a few even bounce straight back), which means there are solid parts to the material. So without ever being able to see directly what is inside materials, we can conclude that materials are made of particles with spaces between them.

Of course the actual experiments that scientists conduct aren't quite as simple as firing marbles at a hidden target. The particles that scientists fire at materials are themselves too small to see. How do scientists then know that they're really firing particles at the materials? They know because the results of such experiments are consistent with the evidence they have gained from other experiments. In these other experiments, they have burned, weighed, smashed, and dissolved materials. They have determined how materials respond to electricity and magnets. The one thing they can't do is look directly inside a material to see whether or not it is really made up of particles. Yet the results of all the experiments scientists *can* do are consistent with the idea that materials are made up of particles.

Scientific Models

Scientists call the idea that all materials are made up of particles the particle theory. A **theory** is an explanation for a set of observations or a group of phenomena. There is so much evidence in favor of the particle theory that it is the idea most scientists accept as an explanation for the behavior of materials or why things are different. We can use this theory to explain many of the observations of materials. So scientists accept the existence of particles and now spend a great deal of time trying to determine what a typical particle might look like.

In the days when scientists first started firing particles at materials and concluding that materials were made of tiny, unseen particles, they had no idea what such a particle might look like. From their observations they could only guess what a particle looked like. Today we have powerful microscopes that allow us to get a picture of some of these particles. The picture still is not clear enough to know for sure what all the parts or components of individual particles look like or even what the inside of a particle looks like. Scientists still perform experiments on materials and use their observations to make their best estimate of exactly what particles look like, inside and out. Scientists call this type of an estimate a **scientific model,** because it is a representation of something they cannot accurately observe directly. They talk about the particle theory and their explanation of what a particle might look like as the **particle model.**

You probably were familiar with many scientific models before this reading. If you remember that a scientific model is a representation of something you cannot easily or accurately observe directly, you probably can think of some of those models. In fact you probably have even created a few of your own models. An example was when you drew and described representations of what you thought was under the cardboard in the mystery box. That is considered a scientific model because you had to describe something that you could never see, but for which you could collect evidence by conducting experiments. You probably also are aware of other familiar models, such as models of body cells and their tiny parts and how they function in our bodies. We can see cells and their parts under a microscope, but using a cell model makes it easier for scientists to explain observations about living things. You also might be familiar with models of the universe, which include stars, planets, galaxies, and solar systems. The universe is an example of something so large that we cannot observe all of its components at the same time. Using models of the universe or its parts helps us understand the universe better.

Stop and Discuss

6. Describe another example of a scientific model.

What Makes a Model Scientific?

Models aren't considered scientific models unless they can explain many, if not all, observations you make when conducting related investigations. This is one criterion that makes a model scientific. The reason the particle model is used by most scientists is that it does explain many of the observations concerning the different ways materials act. A model wouldn't be much help if it explained only one or two observations but left many observations unaccounted for.

Stop and Discuss

7. Based on the criterion that a scientific model explains many, if not all, observations, describe whether or not you think your mystery box model was scientific.

When you developed your own model to explain the mystery box, it probably wasn't too hard to come up with observations your model couldn't explain. So what makes accepted scientific models seem so thorough? Are scientists smarter or more creative than you? Are there just a special few people who can produce strong scientific models? A few reasons why accepted scientific models seem so strong are as follows:

- There are many different scientists working on any given scientific model at any one time. The models and theories you read about in books usually represent the combined work of hundreds of people, not just one or two.
- Scientists develop scientific models over a long period of time. When scientists encounter observations their models can't explain, they often change their models to account for the new observations, but make certain they still explain the old observations. When you read about a scientific theory, you rarely are reading people's first ideas on the subject. What you are reading about is a model that has been changed many, many times.
- When you read about scientific theories, you often don't read about the ones that failed. It's kind of like rehearsing for a play or preparing for an athletic performance. When you're practicing, you don't want the world watching. You try to make your performance as good as possible before you show it to everyone.

Do you think the particle model is a scientific model? Does it explain for you why things are different? The following investigation will give you a chance to decide.

SIDELIGHT

Philosophy and Alchemy

People have long tried to explain how and why things are different. Ancient philosophers wrestled with the problems of what different materials are made of, and whether or not something can come from nothing.

Recall that Aristotle said that each "element" had distinct properties. Air was warm and moist, earth was cold and dry. Water was cold and wet, while fire was warm and dry. Fire could become air through the action of heat; air could become water through the action of moistness; water could become earth through the action of coldness; and earth could become fire through the action of dryness. Thus for Aristotle everything in the world resulted from the combination of these elements and their properties.

Building upon Aristotle's ideas, some people thought that by using the right combination of steps and materials, they could create new materials. These people were known as alchemists, and their art was known as alchemy. Alchemists wanted to find a way to change one kind of material (like lead) into another kind of material (like gold). Alchemy is said to be a mix of science and magic, because alchemists believed that it was somehow possible to produce a magical substance, which they called the "philosopher's stone," that would change ordinary metal into gold. No one, however, ever produced the philosopher's stone, although the desire for gold made it possible for many people to be fooled and swindled by crooked alchemists who claimed that they could convert certain materials into gold.

On the other hand, many alchemists were careful scientists. They kept detailed records of their experiments and developed many techniques for separating and preparing materials used today. Alchemists are responsible for inventing many pieces of the equipment that are still in use in many chemical laboratories. Through their experiments alchemists discovered important facts that became the basis for modern chemistry.

INVESTIGATION:
Model Judges

Do you think that everything in the world is made of tiny particles? For this investigation you don't have to be convinced of that. You only have to observe more phenomena and use the particle model to explain your observations. If you can explain all of your observations using the particle model, then you are verifying that the particle model is a valid scientific model.

Working Environment
You will work individually in this investigation at your desk or at a lab station.

Materials

For each student:
Part B—The Balloon Diet
- 1 balloon
- 1 heat source
- 1 large bowl of crushed ice and water
- 1 metric measuring tape

Part C—Red Streamers
- 2 glass beakers, 600-mL
- crushed ice
- cold water
- red food coloring
- hot water
- 1 medicine dropper
- 2 hot pads

Procedure: Part A—Houdini Water

1. Observe the experimental setup that your teacher has assembled.
2. Predict what you think will happen in this experiment.
   Notebook entry: Record your predictions.
3. Observe what happens in the experiment.
   Notebook entry: Record your observations.
4. Participate in the class discussion about the experiment.
   Be prepared to share your ideas with the rest of the class.

Procedure: Part B—The Balloon Diet

1. Obtain all of the materials for this part.
2. Blow up a balloon as full as you can without popping it, then let out all of the air.
3. Repeat step 2.
4. Blow up the balloon a little more than halfway.
   Do not fill the balloon completely with air.
5. Tie a knot in the end of the balloon to seal it tightly.
6. Measure the balloon with the metric measuring tape around the fattest part of the balloon.
   This measurement is called the circumference of the balloon. Figure 12.2 shows you the right way and the wrong way to measure a circumference.
   Notebook entry: Record this measurement.
7. Place the balloon in the bowl of crushed ice and water for 10 minutes.
   Periodically submerge the balloon into the ice water for 1 minute or so. Do this a total of 4 times during a 10-minute period. Ask your teacher whether it is appropriate to work on Part C while you wait.
8. After 10 minutes, remove the balloon and immediately measure its circumference.
   Notebook entry: Record this measurement.

Figure 12.2 ▼

The fattest part of the balloon is not necessarily the middle part of the balloon as shown here. Where the fattest part of the balloon is depends on the shape of the balloon. You can determine where the fattest part is by looking or by taking several measurements of different areas of the balloon.

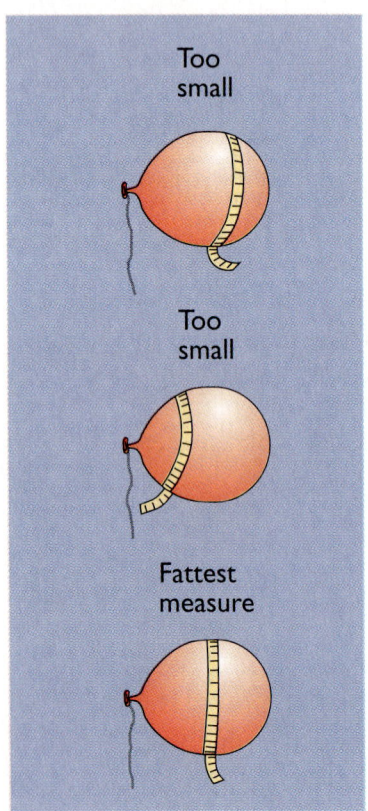

220 ■ Why Are Things Different?   *Elaborate*

9. Place the balloon by, but not touching, a heat source for 10 minutes.

   You will have to ask your teacher what heat source you should use. While you are waiting, ask your teacher whether you should proceed to Part C.

10. After 10 minutes take the balloon away from the heat source and immediately measure its circumference.

    Notebook entry: Record this measurement.

11. Compare the circumferences when the balloon was at room temperature, when the balloon was cold, and when the balloon was hot.

    Notebook entry: Record your observations.

12. Use the particle model to try to explain your observations.

    Explain what is happening with the particles of the balloon and with the particles inside the balloon to make the balloon act the way it does at different temperatures.

    Notebook entry: Record your explanation.

Procedure: Part C—Red Streamers

1. Obtain the materials for this part.
2. Fill one of the beakers with ice and water so the beaker is three-quarters full.
3. Fill the second beaker three-quarters full of hot water.

   Your teacher will provide this hot water when you are ready.

   ▲ **CAUTION:** Use hot pads to handle the beaker containing hot water.

4. Remove all of the ice from the ice water in the first beaker.

   Use a spoon to remove the ice and dispose of it as your teacher directs.

5. Gently drop one drop of red food coloring onto the surface of the water in the middle of each beaker.
6. Observe the action of each drop of food coloring in each beaker.

   Notebook entry: Record your observations.

7. Use the particle model to explain your observations.

   Explain what is different about the particles in hot water from the particles in cold water that results in the different patterns you observed with the red food coloring.

   Notebook entry: Record your explanation.

Wrap Up

Answer the following questions in your notebook. Be prepared to present your particle model explanations for these phenomena to the rest of the class.

1. What was one characteristic that all of the experiments had in common?
2. Decide how scientific the particle model is. Base your decision on the criterion that a scientific model accounts for all observations.

CONNECTIONS:
Particle Movement—Improving the Model

How successful were you in using a particle model to explain what happened in the investigation Model Judges? Did you have trouble? It's perfectly acceptable if you did. It's not easy coming up with models to explain your observations. It has taken scientists quite a few years to develop models that explain many observations, and scientists often need to change models as they make new observations.

In order to help explain your recent observations in the investigation Model Judges, you might need some more information about the currently accepted model for what makes things different. First you can find most materials in one of three forms. A material can be a liquid, solid, or gas. Water, for example, is most commonly found as a liquid. If you boil water, it will turn into a gas. If you freeze water, it will turn into ice, a solid. What makes these three forms of water different is the way the particles are held together. In all materials particles attract each other and have "forces" between them. These forces are stronger in solids than they are in liquids, and in a gas, the forces are even weaker.

Another important part of the model is that particles in materials are in motion. When particles move faster, the material gets hotter. When particles move slower, the material gets cooler. So particles that make up hot water are, on the average, moving faster than particles that make up cold water. Likewise hot air particles move faster than cold air particles. Particles that make up a warm sheet of steel are moving faster than particles that make up a cold sheet of steel.

Answer the following questions in your notebook:

1. Can you now use the particle model to explain your observations in Model Judges? As a class use this new information about the particle theory to explain what happened in the Red Streamers experiment when you put food coloring into hot and cold water. Does this explanation account for all of your observations?
2. Now consider what might be happening with the evaporating liquid in the Houdini Water demonstration during Model

Judges. But before you try to devise an explanation, participate in the class activity your teacher will now conduct.

3. How is what you did in the class activity in question 2 similar to what happened with the particles in Houdini Water?

4. If a model explained that every once in a while particles in the liquid state were not attracted to each other, would that model explain evaporation? Explain your answer.

5. Use all of the information you have about the particle theory to explain what happened to the green food coloring in Houdini Water.

6. Finally, consider what happened in Balloon Diet. Sketch a picture of what happens with the particles inside a balloon at different temperatures.

You have just used three examples to show that a particle model that includes particles in motion can help explain the observations you made in Model Judges. This doesn't mean that the particle model is the *only* model that can explain these observations, though. You also probably could come up with a good explanation using earth, air, water, and fire! In other words, just because the particle model is the currently accepted one (and for good reasons), that doesn't mean it explains everything about the world. For the moment, however, it does an acceptable job of explaining why things are different.

You also should know that we haven't given you all the details of the particle model.

7. To see that the model isn't complete, try to use it to explain the phenomena you observed in the investigation Strange Phenomena. Use the space you left after the explanations you created in the Strange Phenomena Wrap Up to revise those explanations.

Although you might not be able to explain everything, you can use the model to explain quite a few things, as you'll see in the next investigation. You also can use the model as a starting point for developing your own models for what's inside of things, as you'll see in Chapter 13.

INVESTIGATION:
Judging the New and Improved Version

In this investigation you will apply your understanding of the particle model to a new situation. You will gather observations and then rate the particle model as to how scientific it is. You will base your ranking on the criterion that a scientific model explains many, if not all, observations.

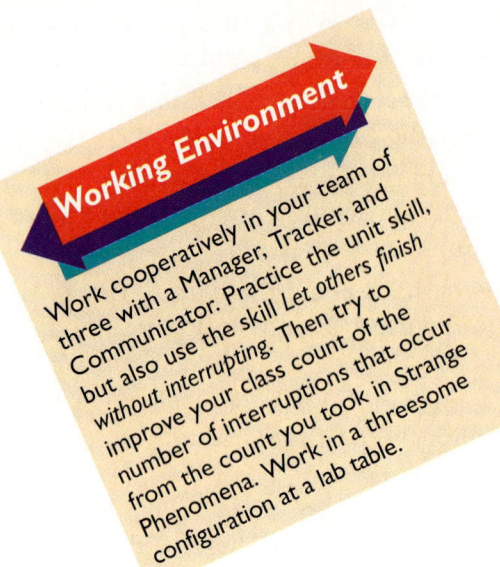

Working Environment

Work cooperatively in your team of three with a Manager, Tracker, and Communicator. Practice the unit skill, but also use the skill Let others finish without interrupting. Then try to improve your class count of the number of interruptions that occur from the count you took in Strange Phenomena. Work in a threesome configuration at a lab table.

Materials

For each team of three students:
Part A—The Amazing Coin Dancers
- 2 glass beakers, 600-mL
- crushed ice
- tap water
- hot water
- 1 glass soft drink bottle, 12-oz
- 1 coin that completely covers the mouth of the soft drink bottle
- 3 pairs of safety goggles
- 1 wax pencil

Part B—Rainmakers
- 1 glass flask with a rounded bottom, 250-mL
- 1 rubber stopper to seal the flask
- 1 glass beaker, 600-mL
- 1 hot plate
- tap water
- ice
- blue food coloring
- 1 medicine dropper
- 2 hot pads or tongs
- 3 pairs of safety goggles

Procedure: Part A—The Amazing Coin Dancers

1. Obtain all of the materials necessary for this part.

   The Tracker should prepare to record the number of interruptions that occur.

   ▲ **CAUTION:** Wear eye protection for this part.

2. Put a soft drink bottle in an empty beaker.
3. Fill the beaker with cold tap water.

   The water should not overflow.

4. Remove the soft drink bottle.
5. Mark the water level in the beaker with a wax pencil.

   Mark the level of water in the beaker after you have removed the soft drink bottle.

6. Empty the beaker, then refill it with hot water up to the line that you marked.

   Your teacher will provide you with this hot water when you are ready.

   ▲ **CAUTION:** Use hot pads to handle the beaker containing hot water.

7. Fill a second beaker with cold water and ice to roughly the same level as marked on the first beaker of water.

   This is called an ice bath.

8. Submerge a soft drink bottle in the ice bath for 2 minutes.

   The bottom of the bottle should touch the bottom of the beaker. The bottle might float so you might have to hold the bottle down for the entire 2 minutes. Don't let any of the ice water get into the bottle.

9. After 2 minutes stick a fingertip into the ice water, and with the moisture on your fingertip, moisten the rim of the bottle by completely encircling it with your wet fingertip.

10. Remove the bottle from the ice bath.

11. Place the coin on the rim of the bottle to completely seal the bottle.

12. Carefully place the cold bottle sealed with a coin into the hot water.

    Do not jostle the coin on the bottle during this process. If you do, reseal the coin with more ice water on the rim.

13. Observe what happens.

    Notebook entry: Record your observations.

14. Use the particle model to explain your observations.

    Explain what is happening with the particles in the water and the particles in whatever is inside the bottle.

    Notebook entry: Record your team's explanation.

    STOP: Remember to practice your unit skill and to keep count of the interruptions.

15. Return the materials.

    Protect your hands with hot pads when disposing of the hot water.

Procedure: Part B—Rainmakers

1. Obtain all of the materials necessary for this part.
2. Put about 300 mL of water in a beaker and bring it to a boil on a hot plate.

   ▲ **CAUTION:** Keep long hair and loose clothing away from the hot plate. Do not attempt to handle a hot beaker without hot pads.

3. Turn off the hot plate, but do not remove the beaker.
4. Fill the empty flask about one-quarter of the way full with very cold water.

   Add some ice to the flask.

Figure 12.3

Add five drops of food coloring and seal tightly with a rubber stopper. The seal must be tight enough so that liquid cannot spill out if you tilt the bottle.

5. Add five drops of blue food coloring to the cold water in the flask and put the rubber stopper in the mouth of the flask (see Figure 12.3).

   If you do not have a rubber stopper, cover the mouth of the flask with a piece of clear plastic wrap.

6. When the boiling in the hot water beaker has stopped, turn the flask on its side and fit the round base over and into the top of the beaker of hot water as shown in Figure 12.4.

7. Observe the water in the beaker for 5 minutes.

   Notebook entry: Record your observations.

Figure 12.4

The bulb of the flask should fit into the top of the beaker. Make sure the water is no longer boiling.

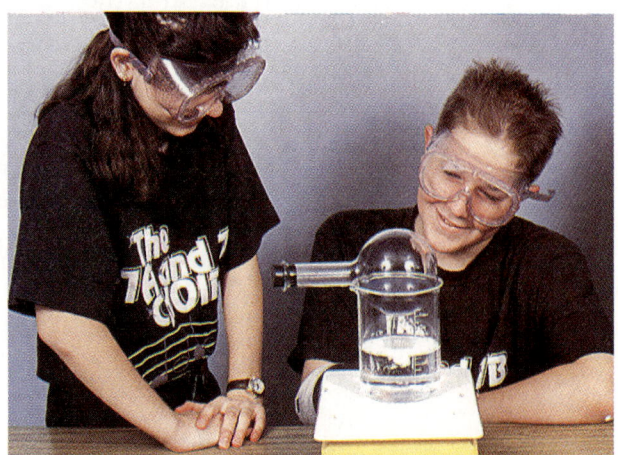

8. Explain your observations by using what you know of the particle model.

   Explain what is happening with the particles in the water in the beaker and the particles in the water in the flask.

   Notebook entry: Record your team's explanation.

9. Return the materials.

   Remember to use hot pads when touching the beaker of hot water.

Figure 12.5

How would you explain the interaction of particles that caused Al's pain?

Wrap Up

1. Share the models your team created for Parts A, B, and C with the rest of the class. Show the class how your model includes ideas of all Team Members.

2. Rate the particle model on a scale of 1 to 15—15 being the best—for how strong a scientific model it is. To do this, assign a score between 1 and 5, 5 being the highest for each of these questions.

   - How well does the model help explain some of your observations?
   - How well does the model explain all of your observations?
   - How complete does the model now seem?
   - The overall ranking is _____. (Add up the scores.)

   The Tracker should write down the rating you gave the particle model along with the reasons you gave the particle model the rating that you did. Have the Communicator hand in to the teacher your team's explanations, your particle model rating, and your reasons for assigning the rating you did.

CHAPTER

13

Using Models to Test and Predict

In the last chapter, you learned a little bit about why people think that all materials are made of tiny particles. You also learned that scientists don't know exactly what these particles look like. They have developed a particle model that, like all scientific models, helps people explain their observations. You learned that one way of answering the question, Why are things different? is to describe what kinds of particles are in various materials and what they are doing. Yet some questions remain.

Think about the Puddle Mystery investigation for a moment. How did the particles from two different liquids come together to make white solid particles that you couldn't see in either of the liquids alone? Think even further back and consider other questions about the properties of materials. What could be going on with particles in materials that make one material hard and another material soft? How could particles in materials be interacting to make one material transparent and another material not transparent? How could the particles be interacting with one another to make one material viscous and another material not very viscous at all?

In this chapter you will have the opportunity to learn more about the particle model. You then will have enough information to answer questions such as these and others that might come up as you develop new explanations for why things are different.

CONNECTIONS:
Model for Sale

The information you have learned so far about the particle theory isn't sufficient for you to answer all the questions in the chapter introduction. You need to work like scientists to try and answer these questions. When scientists have unanswered questions, they often start with an accepted model and revise it. This helps them explain even more observations. You can use this same process.

The particle model you have used so far states that all materials consist of tiny particles and that the hotter the material, the faster the particles move. You also know that particles in solids and liquids attract one another. But exactly *how* do the particles attract one another? What is it that connects particles? Pieces of string? Little springs? And how are the particles arranged? Are they close together or far apart? One way for you to improve the particle model is to specify how the particles are arranged and how they interact with one another.

Here's an example. Watch the paper towel demonstration your teacher will now conduct. Then complete this section by reading and answering the questions in your notebook.

1. Take 5 minutes to develop an explanation that is based on the particle model for the paper towel phenomena you just observed.
2. Share your paper towel explanation with the rest of the class.

Al, too, observed the paper towel demonstration. He used the particle model in a different way to explain what makes paper towels absorbent (see Figure 13.1). His model follows:

Figure 13.1

Al shows the others his arrangement and the interactions of particles to explain why towels are absorbent. He proposes tiny bucket-like structures attached to each particle.

Model for Sale

AL: I've got it! Here's why paper towels hold water! The towel is made up of tiny particles, and each particle has a little bucket-like thing attached to it. When you put the paper towel in water, the buckets fill up. If a towel is really absorbent, that just means the buckets are bigger.

MARIE: I don't buy it, Al. Prove it to me.

AL: See? When I squeeze the towel, water comes out. The towel was full of water.

MARIE: No, you're showing me the observation that led to your explanation. Now put your explanation to the test.

ROS: I have an idea! Here's a way to prove it, Marie. If the particles are attached to little buckets, we should be able to find which end has the buckets pointing up by filling them and quickly flipping the towel around. The water should spill out of the buckets.

MARIE: I'll buy that!

ISAAC: Let's try it. I've got a roll of paper towels and some water!

AL: I'm confused again! (The characters try the experiment.)

ISAAC: Hmmm. Doesn't work. Not a scientific model.

AL: Why not? It explained all *my* observations!

ROS: Well, now there are *more* observations!

AL: Okay, then I'll add something to my model. The bucket things are on swivels and never turn upside-down.

MARIE: Sold!

Al used his knowledge that all materials are made of particles and then added his idea that in paper towels, the particles are attached to bucket-like structures. Al created a new model by using accepted information and his own new ideas.

1. Explain whether or not you think that Al, or any scientist for that matter, should have to test his model in order to convince somebody else.

2. Do you think that Al should have changed his model to fit the new observations that resulted from the test his friends performed on his model? Explain your answer.

Now you know several parts of the particle model: All materials are composed of particles, particles in hot things are moving faster than particles in cold things, particles attract each other, and particles in different materials are uniquely arranged and have unique interactions. The next few investigations will give you the chance to use all of this information to create scientific explanations for more phenomena.

INVESTIGATION: Gloop

In this investigation you will determine some of the properties of a very curious material. Then you will have a chance to develop your own scientific model by using your own ideas and what you already know of the particle theory. Your model will need to explain the properties of the strange material. You also will see what it is like for someone to ask you to test your model and then have to decide how to change your model to fit the results of the tests.

Working Environment

Work cooperatively in your teams of three with a Manager, a Tracker, and a Communicator. Continue practicing the Unit 3 skill, but also try using the new skill Look at the person speaking to you. Work in your threesome configuration at your desks or at a table.

Materials

For each team of three students:
- 3 samples of gloop
- any measuring devices you think you might need

Procedure: Part A—The Social Skill

1. Discuss why the social skill of looking at the person speaking to you is important when working with others.
2. Think of three strategies or things you will say to remind each other to use this skill.

   Notebook entry: Record your ideas.

Procedure: Part B—Gloop's Properties

1. Obtain the materials.
2. As a team explore the gloop.

   Handle the gloop, look at it, pull it, push it, and manipulate it in as many ways as you can. Spend 10 minutes just exploring the gloop.

3. List all Team Members' observations about the properties of gloop in a list titled "Properties of Gloop."

   Notebook entry: Make this list in your notebook. Include all of the observations you made about how it feels, what it looks like, and what it does.

4. Present your list of gloop's properties to the rest of the class.

Procedure: Part C—Explaining the Properties

1. As a team conduct a brainstorming session to find possible scientific models that would explain all of the properties of gloop.

   Review How to #3, How to Have a Brainstorming Session.

2. Create one scientific model that explains all of the properties of gloop.

   Base your explanation on what you know about the particle model and add your own ideas about the arrangement of the particles. Your model should explain why gloop acts, feels, and looks the way it does. Remember, you may create only one model that explains all the properties of gloop that you observed. Record your team's gloop model in your notebook.

Procedure: Part D—Putting Your Model to the Test

Figure 13.2

Notice how the characters base their predictions directly on the model they created to explain gloop's properties.

1. Read through Part D.
2. Construct a data table for recording the information you will obtain as you follow this procedure.

   Notebook entry: Construct this data table in your notebook. Be sure you make space to record the properties of gloop, experiments on gloop, each prediction you make, results of experiments, and what your revised model states after each experiment.

Explore   Using Models to Test and Predict

3. Based on your team's model, predict what will happen if you freeze gloop.

   This experiment is a way of testing how well your team's model can account for new observations.

   Your prediction should answer the following questions:
   - *Will freezing the gloop change its properties that you have noted already in Part B?*
   - *If so, how will its properties change?*

4. Tell your teacher that you are ready to conduct an experiment on frozen gloop, and he or she will provide you with a sample.

   The Communicator should do this.

5. Carefully examine the frozen gloop.

   Notice any new or different properties that you did not observe in your first exploration of gloop during Part B.

   Notebook entry: Record the properties of frozen gloop in your data table.

6. Decide whether your gloop model accounts for the properties of frozen gloop.

7. If necessary, revise, add to, or change your model so it explains not only the properties you observed in Part B, but also the new properties you observed in Part D, step 5.

   Notebook entry: Record any revisions to your model in your data table.

Figure 13.3

Do you agree with Marie, or should the characters revise their model?

STOP! Remember to include everyone's ideas in your revised model.

8. Based on your new model, predict what will happen if you left the gloop uncovered for 48 hours.

   Notebook entry: Record your prediction in your data table. Your prediction should answer the same questions as before.

   - *Will leaving the gloop out for 48 hours change the properties of gloop you listed?*
   - *If so, how will its properties change?*

Figure 13.4

Ros wants to get on with the experiment and not make any predictions. What would you tell her?

9. Tell your teacher that you are ready to conduct the experiment with the gloop that has been left uncovered, and he or she will provide you with a sample.

10. Explore this sample for any new or different properties.

    Notebook entry: Record in your data table the properties of the gloop that has been left uncovered.

11. Decide whether your revised gloop model accounts for the properties of gloop that has been left uncovered for 48 hours.

12. If necessary, revise and add to or change your model so it explains all of the properties you observed in gloop, frozen gloop, and gloop that has been left uncovered.

    Notebook entry: Record your revised model in your data table.

    STOP: Are you looking at the person who is speaking? If you need to make better use of this skill, refer to the three strategies your team agreed to use.

Explore — Using Models to Test and Predict

Figure 13.5

Make sure, as Ros is doing, that your current model accounts for all of your observations of gloop, not just your recent observations of gloop that has been left uncovered.

Figure 13.6

Now what would you say to Isaac? Would you agree or disagree?

13. Predict what would happen to a sample of gloop left soaking in water for 24 hours.

   Notebook entry: Record your prediction in your data table. Your prediction must answer the same questions as before.

   - *Will soaking the gloop in water for 24 hours change the properties you listed in the original gloop sample?*
   - *If so, how will its properties change?*

236 ■ Why Are Things Different?

14. Tell your teacher that you are ready for a sample of gloop that has been soaking in water.
15. Examine the sample for new or different properties from those you observed in your original gloop sample.

    Notebook entry: Record your team's new observations in your data table.

16. Decide whether your current gloop model accounts for the properties of the gloop that has been soaking in water for 24 hours.
17. If necessary, revise, add to, or change your current model so it explains all of the properties you observed in gloop, frozen gloop, gloop that has been left uncovered, and gloop that has been soaked in water.

    Notebook entry: Record your revised model.

Figure 13.7

If you get stuck, make use of the communication system in cooperative learning.

18. Return all materials, including all gloop samples, to the appropriate location.
19. Share your team's original gloop model and your team's revised gloop models with the rest of the class.

    Be sure you tell the class what observations led you to create and revise your model as you did.

Wrap Up

As a team discuss the following questions. Then record answers in your notebooks.

1. Describe how similar or different the teams' models were to one another.
2. Propose a reason for why the models might have been similar.

Explore — Using Models to Test and Predict

3. Propose a reason for why the models might have been different.

4. Pretend that it is 20 years from now. You have just been awarded an International Science and Technology award for your work on a mysterious substance known to the scientific community as "gloop." Your team is at the awards banquet at which you all are expected to give one speech. The prize committee has asked to you explain your modeling process for gloop in your speech. They specifically ask that you tell the assembled scientists how your model changed, how many times you had to revise your model, what caused you to revise your model each time that you did, and how your model includes ideas from each of you. Write your speech in your notebook.

READING:
More On Models

Before you began the previous investigation, Gloop, we asked you a very important question: Do you think that Al, or any other scientist for that matter, should have to test his model in order to convince somebody else?

Stop and Discuss

1. Now that you have created your own models in the gloop investigation and we asked you to test them, how would you answer that question?

Creating a model is hard enough. When someone tells you that you have to *test* your model, the job of modeling becomes even harder. In the scene in which Al had devised a paper towel model, it was easy to be sympathetic with Al and wonder why he had to test his model in order to prove it. It was, after all, *his* model and nobody else had to believe it, right?

Well, that's true. No one *has* to believe any model is right. But in order to be a *scientific* model, other people should be able to use it to predict unplanned observations. In Chapter 12 you learned that one criterion of a scientific model is that it should explain many, if not all, observations. Another criterion of a scientific model is that we should be able to use it to make successful predictions.

Stop and Discuss

2. Outline the basic steps you took when you created models for gloop.

When scientists create models, they think of experiments (or tests) they can do that will result in new observations. Before they conduct these experiments, though, they try to predict what will happen based on their model. Scientists then conduct the experiments and see whether or not they predicted correctly. If they did make correct predictions, they feel that their model is fairly scientific. If the new observations do not match the predictions they made, however, the scientists revise their model. This means scientists change parts of their model so that it explains the new as well as the previous observations. In either case (if the predictions are correct, or if the predictions are incorrect and scientists have to revise the model) they will think of new experiments and conduct more tests in order to make new predictions. Thus scientists spend a lot of time trying to improve their model so that it explains many observations. They then feel confident that their model is as scientific as it can be.

Making Predictions

Making predictions from a model is more than just thinking about what might happen as the result of a certain test. To make predictions you need to make sure that you *base your prediction on your model*. You also need to keep accurate records of the predictions you make. One way of accomplishing these goals is by using if–then statements. An if–then statement first states what your model is and then states your prediction based on that model. Let's look at some examples.

Isaac thought that white construction paper was more translucent than black construction paper. His model for what was going on inside the two kinds of construction paper is shown in Figure 13.8.

Figure 13.8

Isaac decided that all construction paper is made of long chains of particles. In white construction paper, these chains are lined up, and they let light pass through. But in the black construction paper, the chains cross over one another and don't let much light pass through.

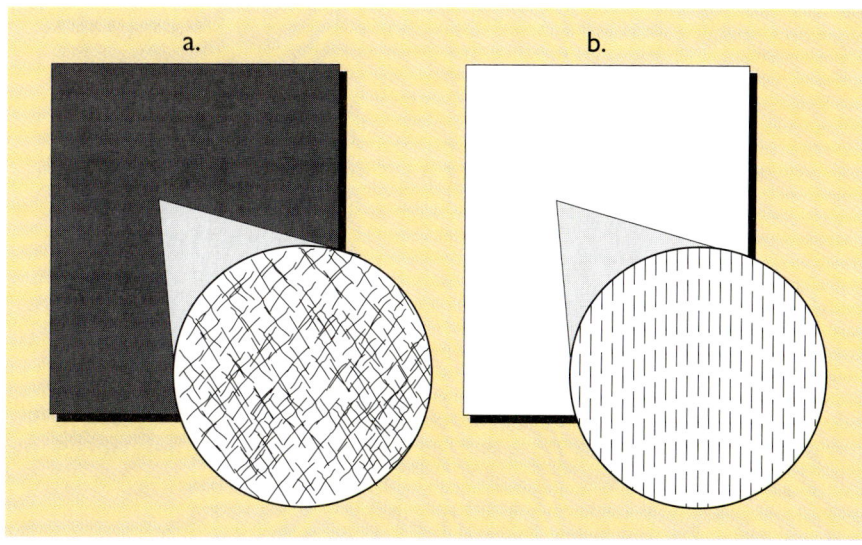

Explain

Isaac used this model to make a prediction. Notice how he uses an if–then statement:

"*If* the particles in white construction paper are all lined up and the particles in black construction paper cross over each other, *then* when I tear the construction paper, the white construction paper should tear as evenly as if I'd cut it with scissors, but the black construction paper should be raggedy."

Notice that the *if* portion states the model, and the *then* portion states the prediction. In this if–then statement, Isaac predicted the outcome of an experiment by basing his prediction on his model. Isaac now has to perform the experiment to see if his prediction was correct.

Stop and Discuss

3. Test Isaac's prediction. Is it correct?

4. If Isaac could successfully predict how the different papers tear, then he would be on his way to developing a scientific model. Would you say Isaac's model is on its way to being scientific?

Consider again the character scene at the beginning of this chapter. If Ros would have used an if–then statement, she might have said:

"*If* paper towels are made of particles attached to buckets, *then* I should be able to fill up the buckets with water, turn the towel over, and see streams of water pouring down."

When Rosalind and Isaac conducted the experiment, they found that this prediction was incorrect. Al then appropriately revised his model to account for the new observations. He didn't create a new model; he only added to the old model.

Stop and Discuss

5. How did Al revise his model?

We can examine one final example by studying how Marie explains translucence. Her model says that it depends on how close or how far apart the particles are in the material. The funny thing is that Al came up with the exact same model to explain the viscosity of corn syrup. He said that viscous materials have particles that are close together, and nonviscous materials have particles that are far apart. Al and Marie put their heads together and came up with this if–then statement to make some predictions.

"*If* nontranslucent materials have particles that are close together and *if* viscous materials have particles that are close together, *then* viscous materials should not be translucent."

The characters managed to make a prediction based on both models!

Stop and Discuss

6. Do you think Al's or Marie's models needs revising? Why or why not?

7. Rewrite the three predictions you made for gloop into if–then statements.

Science Versus Nonscience

We can make a distinction between what is science and what is not science. One important characteristic for something to have in order to be considered science is that it is testable. If you are creating any other kinds of models, you don't have to worry about testability. An ordinary model doesn't have to be testable. But wait! If you are creating models for science, you do have to worry about testability. Anything you want others to consider as science must be testable.

When you create a model, ask yourself these questions:

- Does my model involve elements of magic or superstition?
- Does my model require people to believe something without evidence or "just because" or require them to "have faith"?
- Is there something, besides lack of technology, about my model that makes it impossible to test?

If your answer to any of these questions is yes, your model is not necessarily bad, but it is not considered as science because it doesn't have the important characteristic of testability. There are lots of models that are not considered as science because people cannot test them. Here is an example of a model that people can't test.

> Every material on earth is made of different, tiny, invisible alien creatures from different planets all over the universe. Each kind of material has aliens in it that are holding tiny pieces of the ground from their planet. Each piece of ground is one particle of material. The aliens all stand side by side holding up their pieces of ground from their planet with one hand, and holding each other's hand with a free hand. (Aliens usually have more than two hands.) When there are enough of the alien creatures standing side by side and holding hands, you see a piece of material.
>
> For example, there is a planet one million light years away called Zora. The ground of the planet feels soft, puffy, and hairy. The color of the ground is white. The aliens from this planet, known as Zorites, migrate to earth on a regular basis and squeeze into plants that are common in the southern United States. Inside the plants several thousand aliens stand side by side holding up a piece of fluffy, white ground from Zora with one hand. With the other hand, they

loosely grasp the free hand of the Zorite beside them. (Zorites have three hands.) When a farmer comes around to pick the plant, he sees small puffs of white, which he picks and puts into the back of a big pickup truck. At the end of a long day, he is proud of having harvested a big load of what he calls cotton, which is really aliens from the planet Zora standing side by side, holding hands and pieces of Zoran ground. Each piece of ground they are holding is just one particle of what the farmer calls cotton.

The farmer ships big groups of Zorites all over the world. Some is processed and stuffed into bottles of vitamins, or into people's ears. Some is sent to spinners and weavers. The minute the Zorites feel the spinning process begin, they pop out hook-locks from their hands. These hook-locks interlock and help keep the Zorites from separating so that others can spin and weave them. We will never see these aliens, though, because they set up a force field around their tiny little bodies any time they feel that they are being studied. It is impossible to penetrate this force field with anything. This is how it is with all materials, only other materials are made of creatures from planets other than Zora. Creatures from the planet Golee hold up pieces of Golee ground and link hands with springs. Golee ground is hard and bouncy. Golleans make rubber.

This model is not considered as science because it includes aliens that we have never heard of or planets that we have never discovered. It is not considered as science because there is no way we could develop an experiment to test the model. Why? Because according to this model, the aliens would "sense" we were trying to prove their existence, and they would set up force fields that nothing can penetrate. This does not mean that this model is wrong. It only means that it isn't considered as science.

Stop and Discuss

8. What are other examples of models that would not be considered as science?

9. Explain whether you could consider your gloop model as science or as nonscience.

10. Explain whether you would consider the characters' gloop model as science or as nonscience.

11. Complete this statement: I would rate my team's gloop model as scientific/nonscientific for all of these reasons: _____ .

INVESTIGATION:
Leak-Free Models

In this investigation you again will create a scientific model that is based on the particle model. You will use the model you create to make predictions and carry out different experiments. You will state your predictions with if–then statements as we described in the previous reading.

Materials

For each team of three students:
- 1 plastic vial
- 1 nylon mesh screen, 2-by-3 in.
- 1 resealable plastic bag
- 1 newly sharpened pencil
- a water supply
- 1 piece of cheesecloth
- 1 sink or tub
- 1 pen
- any other materials needed to test your model

Procedure: Part A—Look Ma, No Hands!

1. Obtain a plastic vial and a piece of nylon mesh screen.
2. Fill the plastic vial with water.

   The Communicator should do this.

3. Cover the vial with the piece of mesh screen.

   The Tracker should do this.

4. Take the vial of water covered with the screen and place the palm of one hand on top of the mesh screen.

   The Communicator should do this over a sink or tub. See Figure 13.9a.

5. Quickly turn the vial upside down keeping the palm of one hand on the screen and using the other hand to flip the vial over.

   The Communicator should do this over a sink or tub as in Figure 13.9b.

6. Pull your hand away from the screen without removing the screen.

   The Communicator should do this over a sink or tub as in Figure 13.9c. You must hold the vial straight down and not tilted to one side.

Working Environment

Work cooperatively in your team of three using the roles of Manager, Tracker, and Communicator. You need a work space by a sink or tub. The unit skill and the skill Look at the person speaking to you are both important as you create models to explain phenomena.

Figure 13.9 (a, b, c)

Be sure to perform these steps over a sink or tub.

7. Observe what happens.

   Notebook entry: Record your observations.

8. Create a model that explains what you observed.

   In your model include information about how the particles inside the water, the screen, and in the air interact; how they are arranged; and how they might be attracting each other. Also make sure that your model is testable.

   Notebook entry: Record your model.

9. Read the following test:

   Replace the screen with a piece of cheesecloth of the same size.

   This is a test you will subject your model to.

10. Predict what will happen in the test by completing the following statement. If _____, then when we invert the vial of water using cheesecloth instead of screen, the water in the vial will _____.

    Notebook entry: Write this statement and fill in the first blank with what your model states. Fill in the last blank with your prediction. (For example, will the water spill out?) Remember that you must base your prediction on the model you created for this phenomenon.

11. Test your model as described in step 9 and observe what happens to the water.

    Notebook entry: Record your observations and describe whether or not your prediction was correct.

12. If you based your prediction on your model and it was correct, you do not need to revise your model, and your team can proceed to step 13. If your prediction was not correct, discuss as a team how you could change your current model to account for any new observations.

    Notebook entry: Record your revised model.

13. Think of a second test of your model.

    The second test is up to you. Look over the materials that your teacher has provided and send your Communicator around to get hints from other teams for another test you could perform.

    Notebook entry: Record the test your team designs.

14. Predict what will happen in the test by completing this statement in your notebook;

    If _____, then when we _____, the water will _____.

    The first blank space is for you to describe your model. The second blank space is for you to describe your test. The third blank is for you to tell what will happen as a result of your experiment based on what your model says the particles are doing.

    Notebook entry: Record the completed statement.

15. Conduct your test and determine whether or not your prediction was correct.

   Notebook entry: Record your observations and whether or not your prediction was correct.

16. If your prediction was correct and you based it on your model, you do not need to revise your model at this point, and your team can proceed to Part B. If your model did not successfully predict what would happen in your test, discuss with your teammates how you could change your model to account for the new observations.

   Notebook entry: Record your revised model.

Procedure: Part B—Porcupine Water Bags

1. Obtain a resealable plastic bag and a newly sharpened pencil.
2. Use a scale of 1 to 10 (10 being the highest) to rate yourselves so far in using the unit skill and in looking at the person who is speaking.
3. Fill the plastic bag half full with water.
   The Communicator should do this.
4. Seal the bag shut.
   The Communicator should do this.
5. Beginning with the newly sharpened end of the pencil, poke the bag of water below the waterline and continue pushing the pencil through the bag until the point comes out the opposite side.

   The Tracker should do this holding the bag over a tub or sink. The eraser end of the pencil should be sticking out of one side of the bag, the middle of the pencil should be inside the bag submerged in water, and the sharpened end of the pencil should be sticking out of the other end of the bag as shown in Figure 13.10.

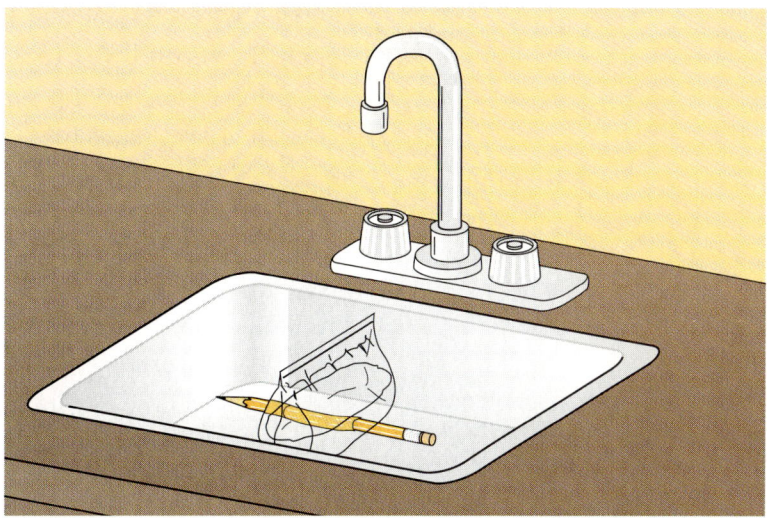

Figure 13.10

Make sure that you are holding the bag over a tub or sink for this step. Notice how the pencil goes straight through the water bag, not at a slant.

Elaborate

6. Observe what happens to the water.

   Notebook entry: Record your observations.

7. Discuss a possible model that explains your observations.

   Be sure your model includes how the particles of the bag, pencil, and water could be arranged and what they could be doing. Be sure your model is testable.

   Notebook entry: Record your model.

8. Read the following test:

   Follow the same procedure except use a pen instead of a pencil to puncture the bag.

   You will subject your model to this test later.

9. Predict what will happen by completing the following statement:

   If _____, then when we *substitute a pen for the pencil*, the water will _____.

   Notebook entry: Write this statement in your notebook and fill in the blanks. The first blank is for you to tell what your model says. The second blank is your prediction. (For example, will the water leak out?) You must base your prediction on your model.

10. Test your model as described in step 8.

    Notebook entry: Record your observations and determine whether or not your prediction was correct.

11. If you based your prediction on your model and the prediction was correct, you do not need to modify your model at this point, and your team can proceed to step 12. If your prediction was not correct, modify your model to account for your new observations.

    Think of other ways the particles could be arranged or other things the particles could be doing that you didn't include in your model.

    Notebook entry: Record your revised model.

12. Think of a second way to test your model.

    The second test you will perform is up to you. Look over the supplies that your teacher provided. This should help you to think of another test. If you still cannot think of another test, ask your Communicator to get ideas from other teams.

    Notebook entry: Record the test you design.

13. Predict what will happen in your second test by completing this statement:

    If _____, then when we _____, the water in the bag will _____.

    The first blank is to record what your model says. The second blank is to describe the experiment you decided to perform. The third blank is to record what you think will happen as a result of your test.

Notebook entry: Record your completed statement.

14. Conduct your second test.

    Notebook entry: Record your observations and whether or not your prediction was correct.

15. If your prediction was correct, you do not need to modify your model, and your team can proceed to the wrap-up section. If your prediction was not correct, revise your model to account for any new observations.

    Notebook entry: Record your revised model.

Wrap Up

Share your results as described in wrap-up question 1, discuss question 2 with your team, and record your answers to question 3 in your notebook.

1. Share your models with the class, explain how you revised your models, and why you revised them as you did. When another team is presenting its model, offer constructive criticism. That means tell them what they did that was good and how they could improve their model even more.

2. Describe how the ranking you assigned your team in Part B, step 2, of the investigation changed.

3. An example of a model for Porcupine Water Bags is that the graphite particles that make up the pencil "lead" do not allow water to seep out. What would you predict would happen, based on this model, if you used a pen instead of a pencil?

INVESTIGATION:
A Penny's Worth of Water

A penny is a fairly small thing, right? What is smaller than a penny? Well, a drop of water is. Have you ever wondered how many drops of water would fit on the surface of a penny? Wonder no more! In this investigation you will have a chance to find out. Then you will have a chance to use all that you know about scientific models to evaluate a model that might explain the phenomenon you will observe.

Working Environment

Work cooperatively in your teams of three with a Manager, a Tracker, and a Communicator. You will choose your own social skill to practice. Work in your threesome configuration at your desks.

Materials

For each team of three students:
- 2 medicine droppers, one for water, one for soap
- 1 beaker, 50-mL, filled with water
- 1 10-mL beaker of liquid soap
- 1 penny
- 2 sheets of paper towels

Procedure: Part A—The Social Skill

1. Review the social skills that you have used this year.
2. Choose a skill to practice from the list that you think would benefit your team. You also may develop a skill of your own.

Notebook entry: Record the skill you choose.

Procedure: Part B—Drops on the Penny

1. Obtain all of the materials.
2. Lay a penny flat on your work area.
3. Have each Team Member guess how many drops of water will fit on the surface of the penny without spilling off.

Notebook entry: Record each teammate's guess.

4. Using the first dropper and the water in the beaker, place drops of water on the surface of the penny, one drop at a time.

   The Manager should do this slowly and patiently without touching the penny or the water on the penny. Be sure to keep the work area stable. Try not to jerk it or bump it suddenly.

5. Help the Manager keep count of how many drops have been placed on the penny by counting each drop aloud as the Manager drops them.

   The Communicator should do this as the Tracker uses tally marks to record the number of drops the Communicator calls out.

6. Immediately stop placing drops of water on the penny when the first bit of water spills over the edge of the penny.

   The Communicator should not count the drop that caused the water to spill off the penny.

Notebook entry: Record the total number of drops that fit on the penny. Notice how close or how far your guesses were from the actual number.

Procedure: Part C—Evaluating a Model

1. Read the following model, which explains why so many drops can fit on the surface of the penny.

   Water particles are extraordinarily lightweight. They are so light, in fact, that if you slowly release water in very small amounts, gravity does not have any effect on the particles. If you release water in large quantities (from a bucket or a faucet for example) there are so many more particles attracting one another that they come out as a heavier mass, and then gravity can have an effect on them.

   If water particles come slowly out of a dropper, however, the drops are small enough that, even though the drop might contain more than one particle, it does not contain enough particles to be affected by gravity. Therefore as you drop water onto a penny, gravity does not pull the water off the penny until there are enough particles of water on the penny to be heavy enough for gravity to pull the water off. When you add the drop of water that makes the rest of the water spill off the penny, that last drop added a few too many particles to the water dome. This makes the water dome heavy enough for gravity to pull it down and off the penny.

2. Discuss this possible model with your team.

   Decide whether it sounds like a scientific model. As you discuss this, consider the following questions:

   - Does the model account for all of the observations you made when you slowly placed drops of water on the penny?
   - Is the model testable? If so, how? What test could you perform on this model? Think carefully! When you have thought of a test, check with your teacher.
   - Try to predict the outcome of the test based on the model. Remember to base your prediction on what the model says should happen, not on what *you* think will happen.

3. Make an if–then statement about the test you would perform and the prediction you would make based on the model by completing the following statement:

   If the model is correct, when we _____, then, based on this model, the result should be (prediction).

   Notebook entry: Record your completed statement.

4. Conduct the test.

   Check your prediction to see whether the model correctly predicted the results.

5. Answer the following questions as a team:
   a. What was the closest guess about how many drops would actually fit on a penny? Were some of you surprised?

b. Can you count the model as science, or is it nonscience?

c. Did the sample model pass the criteria for being a scientific model? In what criteria did it succeed and in what criteria did it fail?

Notebook entry: Record your team's answers.

Procedure: Part D—The Model Needs Revising

1. Read and consider the revised model below.

    Water is composed of tiny particles that constantly attract each other by forces. When you drop water on the penny little by little, the particles in one drop attract not only each other but the particles in the new drop as well. The water particles' attractive forces are strong enough to hold each other together in a dome and are strong enough to overcome the force of gravity to a certain point. Although gravity is pulling constantly on the drops of water, it isn't until the dome of water bulges out just enough that gravity finally can overcome the attractive forces in the water particles and pull the water down off the penny. It is, therefore, the forces between the water particles that are responsible for this phenomenon.

2. Decide whether or not this revised model is a scientific model.

    As you decide, consider these questions:

    - *Does the model explain all of your observations of the water on the penny?*
    - *Is the model testable? How would you test such a model? If it were attractive forces that are responsible for this phenomenon, is there some test you could do that would destroy the attractive forces of the water particles?*
    - *If someone told you about a test that you could do that would break the attractive forces, could you correctly predict the results based on the revised model?*

3. Read the following test, which you will subject this model to:

    To test this model, you will repeat the experiment. But this time you will place fewer drops of water on the penny. Have the Tracker check in his or her notebook for the number of drops that fit on the penny without any water spilling off. Now subtract 10 from that number. This is the number of drops you will place on the penny this time. Then you will place a drop of liquid soap on the dome of water on the penny.

4. Complete the following if–then statement:

    If the model is correct, then after we place a drop of soap on the dome of water, based on this model, the results should be _____.

    Notebook entry: Record the completed statement.

5. Conduct the test as described in step 3.

Remember to use the dropper that goes with the beaker of soap.

Notebook entry: Record your observations.

6. Based on the model, explain what the soap particles did to the water particles that caused them to spill off the penny.

   Your Communicator can check with other groups for possible answers that you did not think of.

   Notebook entry: Record your explanation.

7. Share your explanations with the rest of the class.

Wrap Up

Discuss the following questions with your teammates. Then record your answers in your notebook. Each of you should be prepared to explain your answers in a class discussion.

1. Which of the two criteria for scientific models did the second model meet?
2. How would you rate this revised model as a scientific model: strong, medium, or weak?

READING:
A Promise Is a Promise

The revised model that you just evaluated is actually the current scientific model for the dome of water on the penny that you observed. This model can explain other phenomena as well. Have you ever filled a drinking glass too high and noticed how the liquid bulges over the rim but does not spill over? We also could use the revised model to explain that. The attractive forces between water particles actually have a name. They are called **cohesive forces.** And the phenomenon also has a name. It is called **surface tension.** The cohesive forces between water particles hold water together in this special way.

Let's look at how cohesive forces can help explain another phenomenon you observed. In Unit 2 you spent a lot of time designing boats. The first boat you designed was called a Tom Thumb boat. Remember we promised you an explanation for why the addition of a drop of soap propelled the boat? Well, finally we will keep that promise.

You placed the Tom Thumb boat on the surface of the water. The particles on the water surface below the boat and surrounding the boat were being held together by cohesive forces that resulted in surface tension. There also were forces between the boat's paper particles and the water particles. Those forces are called **adhesive forces.** But the cohesive forces among the water particles are stronger than the adhesive forces among the paper particles and the water particles. So the boat stayed afloat on top of the surface. Then you added a drop of soap to the back of the boat just as you

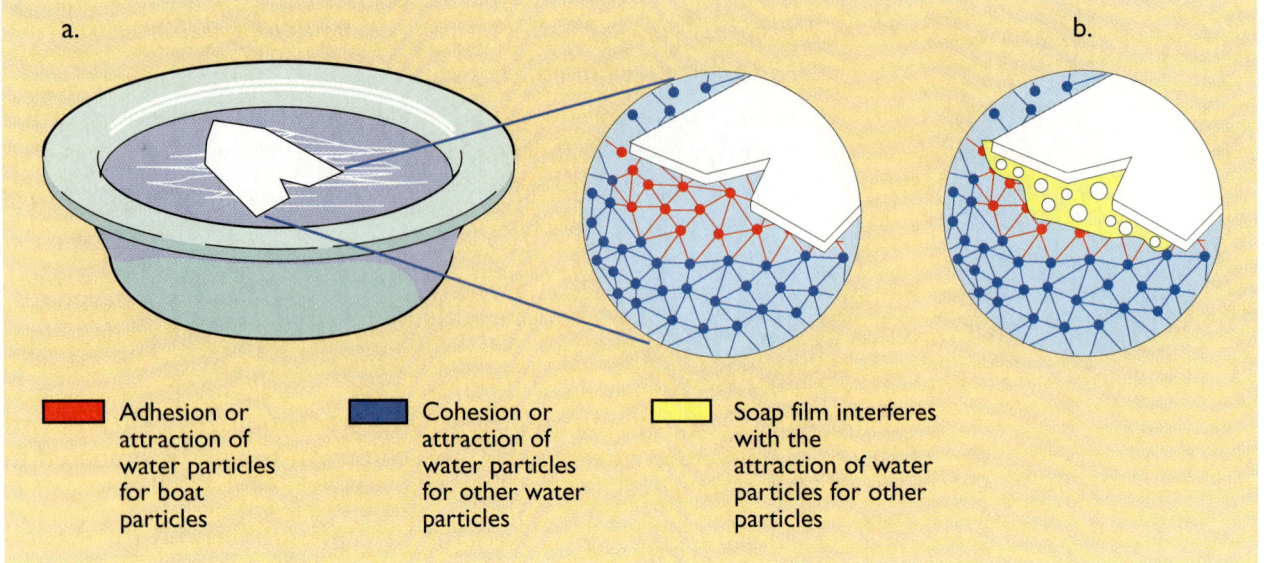

Figure 13.11

(a) First the boat floats on top of the water because the forces between the particles of the paper boat and the forces of the water particles are not as strong as the forces between the water particles themselves. (b) After you add soap, the bouncing movement of particles that are no longer cohesively bonded at the back of the boat overcome the adhesive forces at the sides and front of the boat. The water particles at the back of the boat therefore push the boat along the surface of the other water particles, which are still cohesively bonded.

added a drop of soap to the water on the penny. The addition of the soap caused the boat to suddenly race forward across the water as if it were being pushed. Why?

Think of what happened to the water on the penny after you added the drop of soap. The dome fell off the penny, right? If it were the cohesive forces in the water that held the water in a dome shape, then the soap must have done something to break apart the forces so that the water wouldn't hold together anymore. In fact, the soap particles slipped in between the water particles, blocking the cohesive forces between the particles so that they couldn't hold together any longer.

While the Tom Thumb boat was floating on top of the water, it was surrounded by water that was held together by cohesive forces between the particles. The cohesive forces of the water particles also pulled on the boat, as in Figure 13.11. When you added the drop of soap to the water at the back of the boat, the soap slipped between the water particles in that area and blocked the cohesive forces. Now remember that particles in liquids move about. When the cohesive forces in water particles at the back of the boat were gone, the particles could bounce around more freely, even bouncing against the back of the boat. The rest of the particles to the sides and front of the boat were still held together by cohesive forces, but the particles at the back exerted pressure at the back of the boat, pushing it across the top of the water. The rest of the water particles were still bonded cohesively. Study Figure 13.11 to see a diagram model of this particle phenomenon.

We say that soap broke the surface tension at the back of the boat, just as it broke the surface tension of the water on top of the penny. We have solved two mysteries by one model at the same time. Models definitely have their place in science.

CONNECTIONS:
Properties and Models in Review

Work by yourself to see what you have learned about properties and models. Feel free to turn back in your science book or in your notebook to help you remember the ideas you need for answering these questions.

1. If a new student came into your classroom and it was up to you to define *properties of materials* for the student, what would you say?
2. Do you think materials like the canvas of a backpack, the paper in a book, or the metal of a locker can have insides? If so, what do you mean by "insides"?
3. What evidence do scientists have for the particle model?
4. What evidence did the Chinese and Greeks have that led them to their ideas about why things are different?
5. Are today's scientists smarter than the ancient Chinese or Greeks?
6. Imagine that you are still explaining things to a new student in your class. Explain what a scientific model is.
7. Now explain the particle model to the new person in class.

In Chapter 12 you used your knowledge of the particle model to create your own models for why things had certain properties. By then you had a fairly good idea about what things make a model scientific. The two criteria that determine a scientific model are (1) the model explains many, if not all, observations, and (2) the model can successfully predict new observations. You also learned that in order to be considered science, a model must have the characteristic of testability. Continue your discussion as you answer these questions.

8. Why should a model have to explain as many observations as possible? Why can't a model explain only the first observations you make and then a completely new model explain any new observations?
9. Why do scientists test their models?
10. If you are unsuccessful in accurately predicting new observations based on your model, what should you do and why?

That's a lot of new ideas to learn about and to work with in just one unit. These ideas might not be crystal clear to you right now, but you should be fairly comfortable with these ideas and what they mean. This is the time to ask your teacher and classmates questions and to speak up if you still do not understand something.

It's okay. I know who I am now. I think it's great that I'm a drawing. I'm a miracle of modern scientific technology, and in a way I'm a model because I'm a representation of something you're not sure is there. Heck, I'm not always sure I'm all here! Oh boy, the things I could do in the last unit...

254

UNIT 4

Why Are We Diverse?

In Unit 1 you investigated diversity among your classmates to answer the question, What is normal? The focus question for Unit 2 was, How does technology account for my limits? Then you learned about the design process to help answer that question. In Unit 3 you explored what scientific models are and how you can use them to answer questions. You explored a model that helps answer the question, Why are things different? In this final unit you will have a chance to use what you learned in the previous three units to finally answer the question, Why are we diverse?

First you again will explore the diversity among your classmates. You will then use the scientific modeling techniques you used in Unit 3 to begin developing your own explanation for the reason behind that diversity. After you have studied the current theories about why we are different, you will have a chance to design a book. By the time you are finished with this unit, you should have a fairly good answer to the question, Why are we diverse? You also might have some definite ideas about whether being different from one another is good or bad.

COOPERATIVE LEARNING OVERVIEW

Now that you have been through three units in which you worked cooperatively on many activities, you might feel somewhat accomplished, as Al does. Your ideas about cooperative learning might have changed since the first activity in Unit 1. Can you recall how you felt then about working cooperatively? How has your attitude about cooperative learning changed?

Cooperative learning can mean different things to different people. Marie felt that cooperative learning was all about getting along with others. Isaac felt that cooperative learning had more to do with taking advantage of a variety of people's thought processes to come up with answers. Do you agree with Isaac or with Marie? What does cooperative learning mean to you?

In this unit you will have a few more opportunities to work cooperatively. You will work in teams of two and in combined teams of two, making teams of four. You will not only use the same roles, but you also will continue to practice skills with each activity. With each activity, you will practice the new unit skill of disagreeing with the idea, not the person.

Study the character scene. Why does Ros make the comment that she does? If you could rewrite Isaac's bubble, what would you write so that he would be demonstrating the unit skill? After you have rewritten Isaac's bubble, be sure to meet with your new team and write some of your ideas about ways that you could disagree with each other's ideas but not with each other.

CHAPTER 14

You: A Model for Diversity

If you glance around the room, you probably can identify many ways in which you differ from others. Sometimes you have to know someone well before you realize how different the two of you might be. Take a moment to list in your notebook 10 characteristics that make people different from each other. Then share some of your ideas with your classmates. Why don't we all have the same traits? What is it that makes us different? In this chapter we will explore answers to the question, Why are we diverse?

INVESTIGATION:
Taster's Choice

You know that people differ in many ways. Do you suppose we differ in the way things taste to us? In this investigation you will answer the question, Do all people experience the same taste sensation when they taste paper?

Working Environment
You will work individually at your desk.

Materials
For each student:
- 2 sample strips of paper
- 1 pencil
- 1 pair of scissors

Procedure

1. Obtain the first taste-strip sample.
   Your teacher will hand this to you.
2. Fold the strip in half widthwise.
3. Cut the strip into two pieces.
   See Figure 14.1a.
4. With a pencil label each half piece at one end, as shown in Figure 14.1b.
   Label one half with 1a and your initials. Label the second half with 1b and your initials.
5. Obtain the second taste-strip sample.
6. Fold and cut the strip in half.
7. Label each half with 2a or 2b and your initials.
   Do this as you did in step 4.

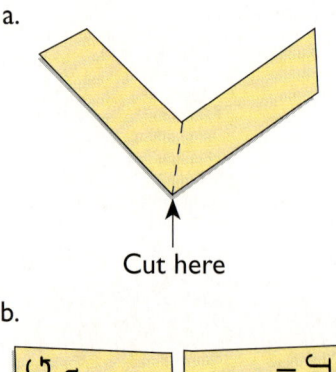

Figure 14.1

After folding the strip in half, cut or tear the halves apart. Then label one end of each strip with your initials and 1a or 1b.

260 ■ Why Are We Diverse? *Engage*

8. Taste pieces 1a and 2a by lightly touching the blank end of each piece to your tongue.

   Taste the end of 1a several times and then the end of 2a several times to make sure you can describe what you are tasting.

9. Describe what each piece tastes like, noting all taste sensations including aftertaste.

   Notebook entry: Record your description.

10. Dispose of strips 1a and 2a in the trash can (see Figure 14.2).

▲ **CAUTION: To help prevent the spread of disease, discard only your own used strips.**

Figure 14.2

Dispose of your own used strips immediately and properly.

11. Save strips 1b and 2b. Do not taste them.
12. If you are a taster, move to the tasters' area of the room.

    Your teacher will let you know which paper strip (the first or second sample) was coated with a special chemical that only some people can taste. This special paper is called thiourea (Thy oh yoo REE uh) paper. If you tasted something on the thiourea strip, you are a taster.

13. If you are not a taster, move to the nontasters' area of the room.
14. Take turns describing to the class what you tasted.
15. Discuss the following with your classmates:

    Explain whether or not you are sure that members of your class experience different taste sensations when they taste thiourea paper.

    What would you need to know before concluding that your classmates taste thiourea paper differently?

16. As a class design an experiment you could perform that could convince somebody else that your classmates taste thiourea paper differently.

Wrap Up

Work on the following challenge individually. Write your response in your notebook.

Think of an explanation for why people in your class differed in their ability to taste the strips of thiourea paper. Make sure your explanation is testable. Be prepared to share your explanation with the rest of the class.

INVESTIGATION: Wheel of Traits!

When scientists describe the characteristics of living things, they call the characteristics **traits.** You and your classmates have thousands of traits. You already know one of your traits: whether or not you can taste thiourea paper. In this investigation you will examine yourself for many traits and answer the question: is your combination of traits unique?

Working Environment

Work cooperatively in your new team of two and use the roles of Manager and Communicator. As you work practice the unit skill and the skill Praise others. Work facing each other by pushing your desks together or sitting across from each other at a table.

Materials

For each team of two students:
- 1 mirror
- 2 pencils with erasers

Procedure

1. Study the photographs in Figure 14.3.

   This figure explains the seven traits you will explore during this investigation.

2. Beginning with the first trait, attached or unattached earlobe, look at the pictures and decide whether your earlobe is attached or unattached.

Seek the opinion of your teammate and use the mirror, if necessary.

Notebook entry: Record whether your earlobes are attached or unattached.

3. Repeat step 2 for your teammate.
4. Go through each trait in Figure 14.3 with your partner and help each other decide whether you have the trait pictured.

   If you are not sure about the trait, or are unsure whether you or your partner have the trait, be sure to read the special notes. If you still need help, have the Communicator check with other teams.

   Notebook entry: Record whether or not you have each trait pictured in Figure 14.3.

Characteristics That Appear on the Trait Wheel

If most of the bottom of your ear hangs free from the side of your jaw, then your lobe is unattached.

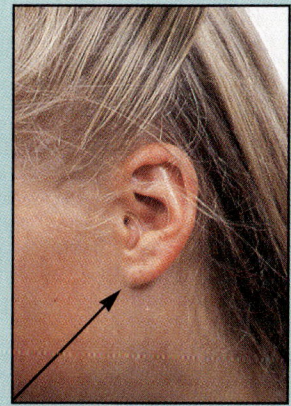
Attached earlobes

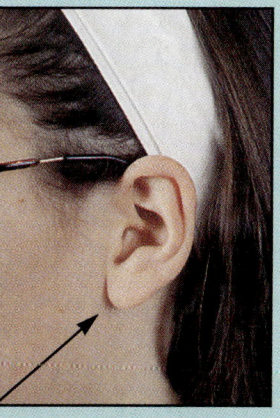
Unattached earlobes

Even one hair on one knuckle means you do have mid-digital hair. Look for signs of worn-off knuckle hairs. Even these count as mid-digital hair.

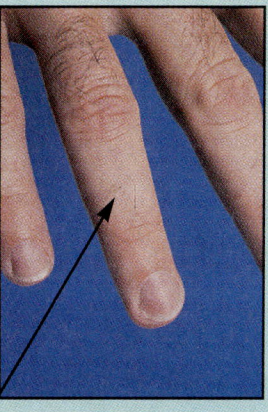
Mid-digital hair

Figure 14.3

You will refer to these photographs to help you determine your traits.

This trait appears as crooked or bent fingers. Even one finger that is bent slightly means you have camptodactyly. Usually it is the pinky that is bent inward.

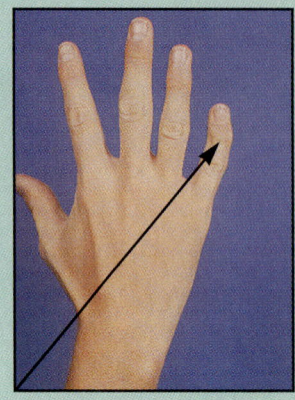
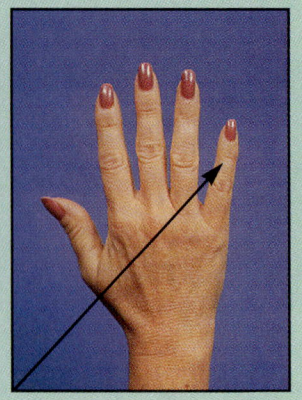

Camptodactyly   No camptodactyly

Just one of these bumps in your mouth means that you have this trait, even if it is small.

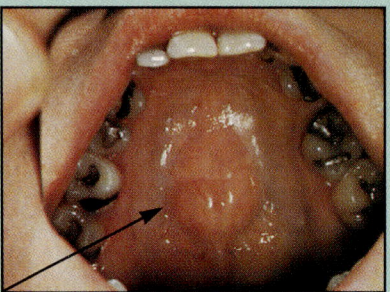
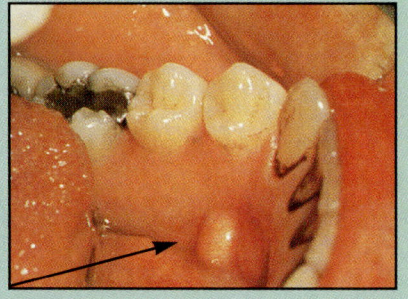

Bony Tori roof of mouth   Bony Tori under tongue

This trait is for facial freckles only, not for ones on the shoulders, chest, or other places on the body. Look on the upper cheek under the eyes and on the nose. If you see only one freckle, you don't have freckles.

Freckles

Look for even a slight dimple in the middle of the chin.

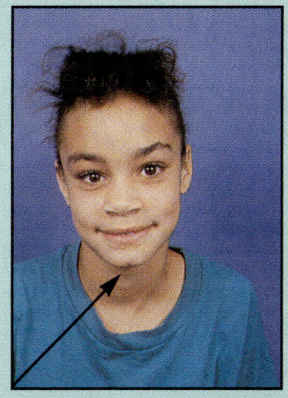

Cleft chin

If the back of your thumb forms a smooth curve with no protruding bones when you bend it back, then you have a hitchhiker's thumb.

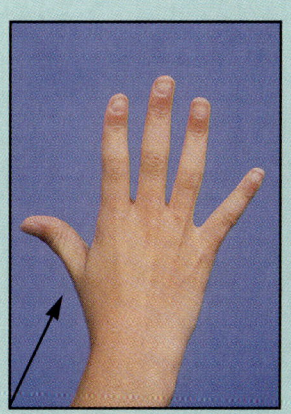 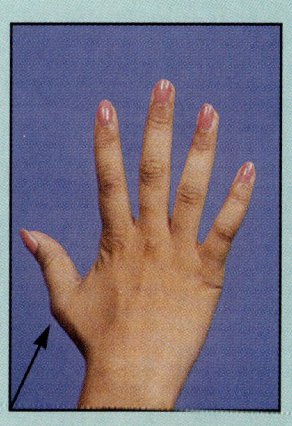

Hitchhiker's thumb — No hitchhiker's thumb

Engage ■ *Explore*

You: A Model for Diversity ■ **265**

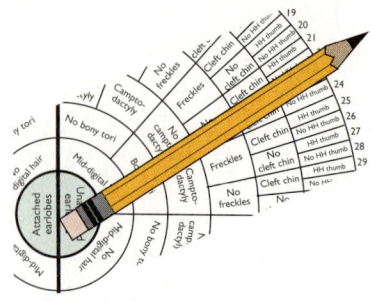

Figure 14.4

This is the first section of the Wheel of Traits. If you have attached earlobes, place your eraser in the space for attached earlobes. If you have unattached earlobes, your pencil eraser should be in the space as shown.

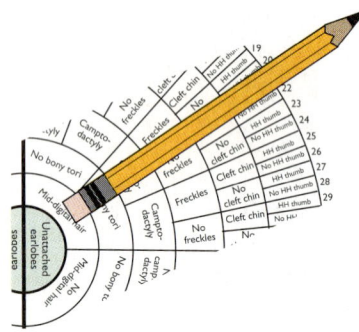

Figure 14.5

From your pencil eraser's position for attached or unattached earlobes, slide your pencil eraser to the section for mid-digital or no mid-digital hair, according to whether or not you have mid-digital hair.

5. Study Figure 14.6 located at the end of this procedure.

   This is the Wheel of Traits.

6. Hold your pencil upside down and place your pencil eraser in the center of the circle.

   Work together as you do this.

7. Slide your pencil eraser to the space that matches your trait: attached or unattached earlobes.

   Hold your pencil eraser in this place (see Figure 14.4).

8. Slide your pencil eraser to the space for mid-digital hair or no mid-digital hair, according to your records.

   See Figure 14.5.

9. Continue moving toward the outside of the circle, sliding your pencil eraser from trait to trait, according to your records.

10. Find the number located just outside the space where you ended on the sheet.

    This number is your trait number.

    Notebook entry: Record your trait number.

11. After both teammates know their trait numbers, record both numbers on the class trait number list.

    Your teacher will tell you where this list is.

Wrap Up

Discuss the following with your teammate and be prepared to share your answers with the rest of the class.

1. How many students are in your class?
2. How many students in your class had the same trait number?
3. Based on the results of this investigation, explain how diverse your classmates are.
4. List several questions you have about people and traits now that you have finished this investigation.
5. Develop an explanation for the diversity of trait numbers in your class.
6. Describe to your teammate an instance in which he or she praised you and how you felt at that time.

Figure 14.6 ▶

Use this Wheel of Traits as indicated in the procedure to determine your trait number. Use your pencil eraser as a guide. If you have questions about what the phrases mean refer to Figure 14.3.

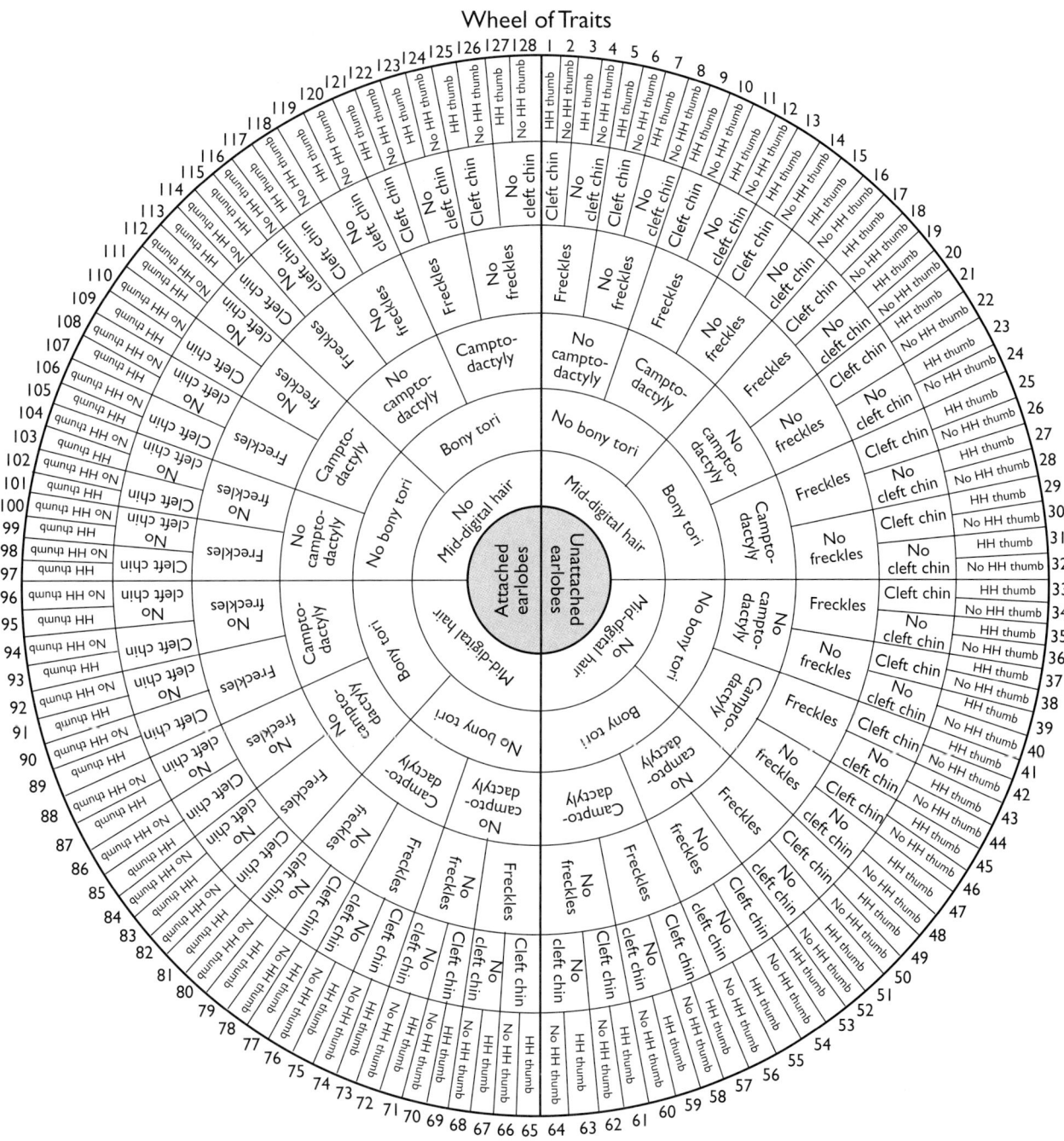

Explore — You: A Model for Diversity — 267

CONNECTIONS:
Traits and Trees

"You have Great Uncle Herbert's nose!" "You look just like your grandmother when you smile!" Have you ever heard comments like these? How similar are your traits to the traits of other people in your family? One way to find out is to review your family history by looking at a **family tree.**

A family tree is a method of recording data about a family for many generations. Some families keep detailed records about their trait history. After several generations they can organize their family trait history as a tree that displays all of the known relatives and their known traits. Sometimes people take special interest in their families and conduct research to learn more about their families. Scrapbooks, photo albums, journals, and relatives are

Figure 14.7

This family tree shows the trait history for four generations of Marie's family. The traits included are cleft chin, hitchhiker's thumb, freckles, mid-digital hair, and bony tori.

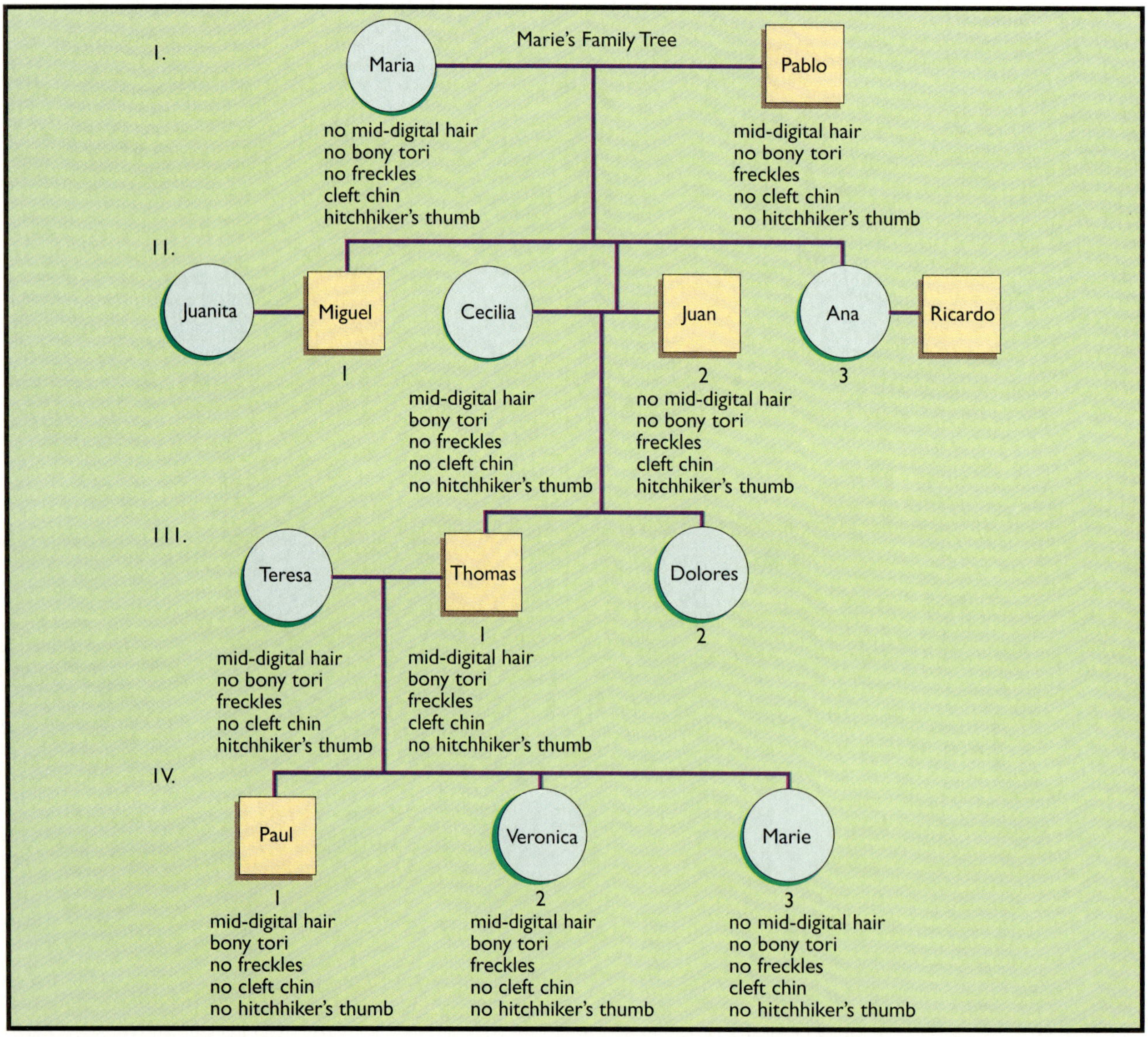

Marie's Family Tree

I.
- Maria: no mid-digital hair, no bony tori, no freckles, cleft chin, hitchhiker's thumb
- Pablo: mid-digital hair, no bony tori, freckles, no cleft chin, no hitchhiker's thumb

II. Juanita — Miguel 1   Cecilia   Juan 2   Ana 3 — Ricardo
- Cecilia: mid-digital hair, bony tori, no freckles, no cleft chin, no hitchhiker's thumb
- Juan: no mid-digital hair, no bony tori, freckles, cleft chin, hitchhiker's thumb

III. Teresa — Thomas 1   Dolores 2
- Teresa: mid-digital hair, no bony tori, freckles, no cleft chin, hitchhiker's thumb
- Thomas: mid-digital hair, bony tori, freckles, cleft chin, no hitchhiker's thumb

IV. Paul 1   Veronica 2   Marie 3
- Paul: mid-digital hair, bony tori, no freckles, no cleft chin, no hitchhiker's thumb
- Veronica: mid-digital hair, bony tori, freckles, no cleft chin, no hitchhiker's thumb
- Marie: no mid-digital hair, no bony tori, no freckles, cleft chin, no hitchhiker's thumb

good sources of information if you want to assemble a tree that shows the trait history of your family.

One thing about family trees is that they are often incomplete. This is because family records are not always kept up-to-date or because someone died and no one living could remember specific information about that person.

Still a family tree offers a wealth of information about common traits in a family. Family trees usually are drawn and written in a special way that you will need to understand before you can study them. After you understand the signs and symbols of a family tree, you will examine Marie's family tree as an example of what a typical one looks like. Then you will determine which of her relatives she resembles the most and develop an explanation for those similarities.

Compare Marie's family tree in Figure 14.7 with the chart in Figure 14.8. This chart is a key to the signs and symbols that are

Figure 14.8

Use this key to help you understand Marie's family tree.

Key for Reading a Family Tree		
Symbol	Description	Meaning
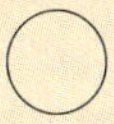 ◯	Circle, has a name in it	• Female
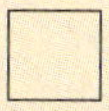 ☐	Square, has a name in it	• Male
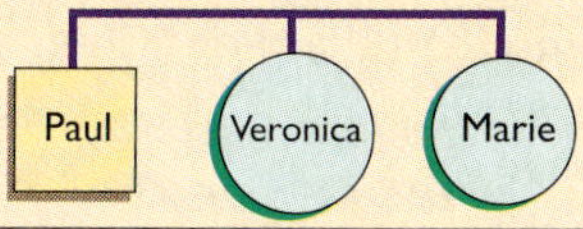 Paul — Veronica — Marie	Horizontal line with names attached by small vertical lines	• Indicates a generation of brothers and sisters. All names attached to the horizontal line are brothers or sisters.
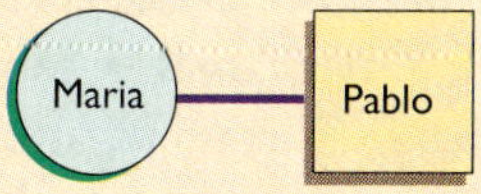 Maria — Pablo	Two names attached only by a horizontal line.	• Indicates a marriage • The in-law has no connection to any other horizontal line. The blood relative is attached to the horizontal brother/sister line.
|	Vertical line dropping down from the horizontal line that connects married people.	• Connects the generations • Shows that this couple had the children that appear on the horizontal line that is attached to the bottom of the vertical line
I, II, III, IV	Roman numerals to the left of each horizontal row of brothers and sisters	• Number of generations shown on the family tree. The lower the number, the older the generation. • Generation I shows the first couple whose traits are known. This couple originates the blood line on this tree.
1, 2, 3	Arabic numerals under a circle or square	• Number the birth order of the children in a family. • 1 indicates the oldest child in the family, 2 the second oldest, and so on.
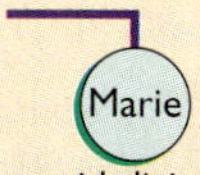 Marie no mid-digital hair no bony tori no freckles cleft chin no hitchhiker's thumb	List of traits under a circle or square	• Indicates the traits that this person was known to express • If no traits are listed, they are unknown for that individual.

Explore

typical of a family tree. Then answer these questions in your notebook.

1. Who are Marie's siblings (brothers and sisters)?
2. What are the names of Marie's parents?
3. Are Teresa, Thomas, and Dolores all siblings?
4. Who are Marie's grandparents?
5. Are Marie's grandparents related to her mother or her father?
6. Marie's great aunts and uncles also are Thomas's and Dolores's aunts and uncles. What are their names?
7. Can you name an aunt and an uncle to whom Thomas and Dolores are not related?
8. Can you name Marie's aunt?
9. What is the relationship between Marie and Maria and Pablo?
10. Who are Maria and Pablo's children?

Now read the following story about Marie.

After Marie did the Wheel of Traits investigation at school, she took it home and tried it with her family. Together they filled in their family traits on their family tree. They used pictures to determine as much as they could about relatives who had died and talked to the living relatives to determine their traits. As her family was doing all this, Marie began to get concerned. She noticed that her father, Veronica, and Paul all had bony tori, but she did not. Then she saw that both she and her father had cleft chins. But then she noticed that neither she nor Paul had freckles, but both of her parents did. She also realized that she was the only one in her family without mid-digital hair. It was then she decided that what she had always wondered about must be true—she must have been adopted! She went to her parents and asked them to tell her the story of her adoption. Instead her parents suggested they all look a little harder at her family tree and see whether that would help answer her questions.

You can help Marie understand why she can have such different traits from her parents and still not be adopted by determining answers to the following questions. Be sure to record the answers in your notebook.

1. Which of Marie's relatives have cleft chins?
2. Which ancestor on this tree first passed on the trait for no mid-digital hair?
3. Explain the fact that Marie doesn't have freckles.
4. Which of Marie's ancestors first displayed and passed along a trait for bony tori?
5. If you could look at the pictures in Marie's family album, who do you think she would look the most like? Explain your answer.

6. Marie has several traits her brother, sister, mother, and father do not have. How can you explain this?

Marie returned to her parents and said, "I get it. I might not look very much like either of you, but I do look like your parents and their parents. I'm still curious, though. *Why* are there traits like that? How come some of us have the same traits all down the line, and others of us have traits that skip generations? How are traits passed down?" Marie's mother provided this explanation: _____. *You* fill in the blank! Try to think of an explanation about how traits pass from person to person.

READING: A Model That Explains Diversity

Marie's questions are good ones. You've just spent a considerable portion of this chapter developing explanations for why people have different traits and how those traits are passed down from one generation to the next. We use the word **inherit** to mean that traits are passed down. To help you elaborate on your explanations about inheritance, you might need some background information. Then you can revise your explanations.

Things You Should Know: Cells and Chromosomes

All living things are made of cells. Cells are microscopic building blocks that make up the body. Cells contain special structures that function to maintain our bodies. Every part of our body is composed of cells: skin, muscles, bone, teeth, blood, and more. Different types of body cells look different and do different things. Study the pictures of cells in Figure 14.9. Notice that there is a part in each cell that is darker than the rest of the cell. This dark part is called the **nucleus** (NOO klee us) of the cell. (The plural of nucleus is **nuclei** (NOO klee eye). The nucleus stores long thread-like strands of material called **chromosomes** (KROH moh sohmz). Chromosomes, in turn, are made up of particles called **deoxyribonucleic** (dee OK sih RY boh noo KLEE ik) **acid**, or **DNA** for short.

Your body has two major types of cells: **body cells** (such as those we mentioned, including skin, muscles, bone, teeth, and blood), and **sex cells.** One of the things that makes these two types of cells different is the number of chromosomes each type contains. Body cells have 46 chromosomes in the nuclei. These 46 chromosomes exist in pairs, so that each body cell has 23 pairs of chromosomes in its nucleus (see Figure 14.10). Sex cells have half the number of chromosomes that body cells have or one set of the 23 pairs found in body cells. The chromosomes in sex cells, therefore, do not exist in pairs. A female's sex cells are called eggs. A male's sex cells are called sperm. Each egg in the female body

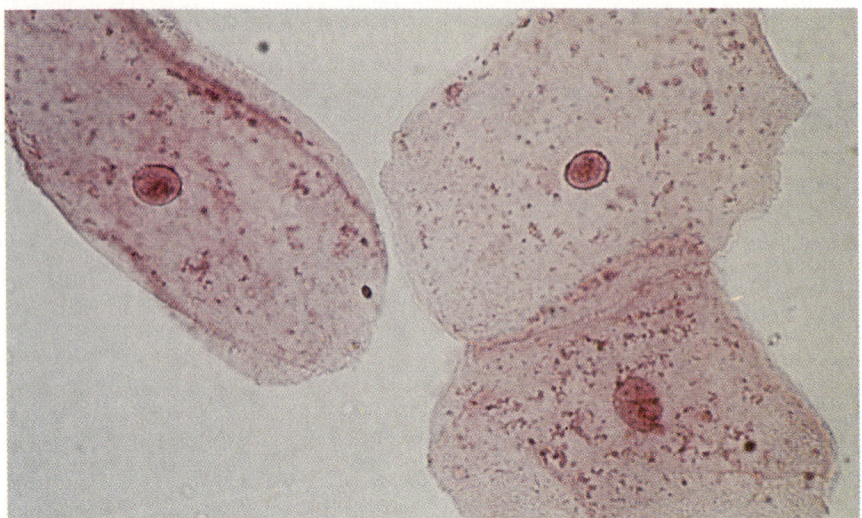
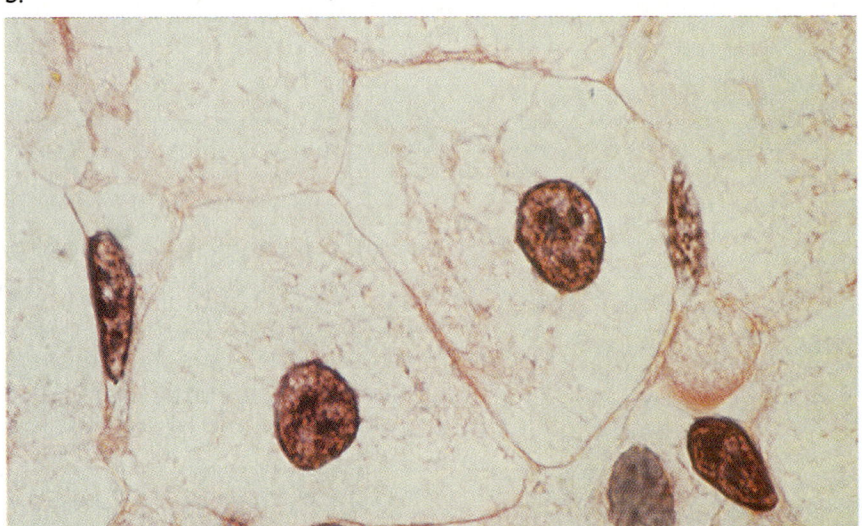
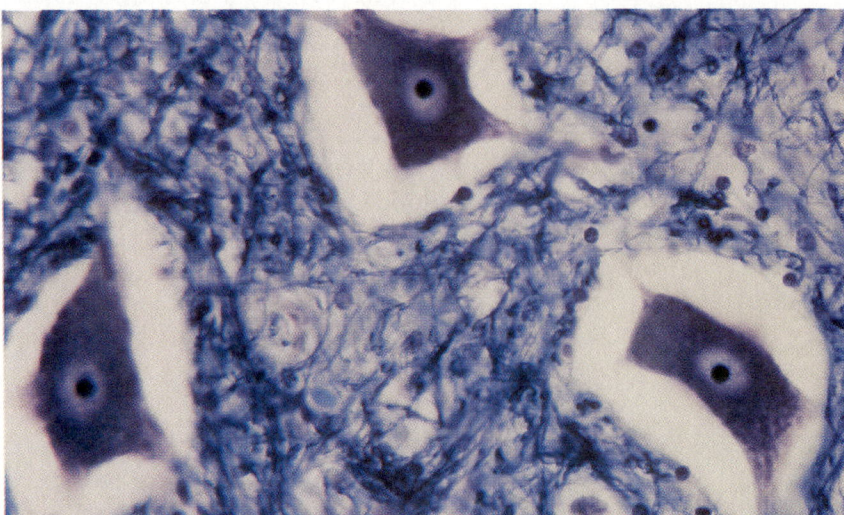

Figure 14.9

These are photographs of (a) human cheek cells, (b) human liver cells, and (c) human nerve cells. The cells are specially treated so you can see some of their special structures, especially the nuclei.

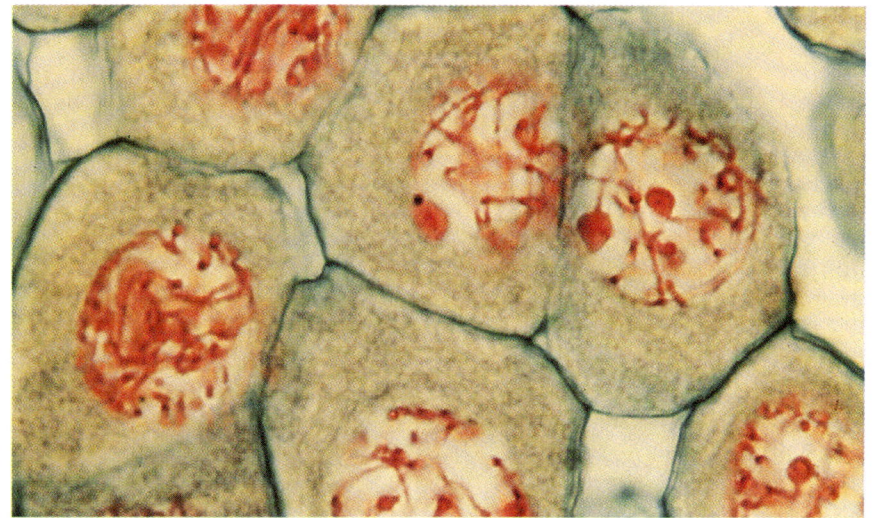
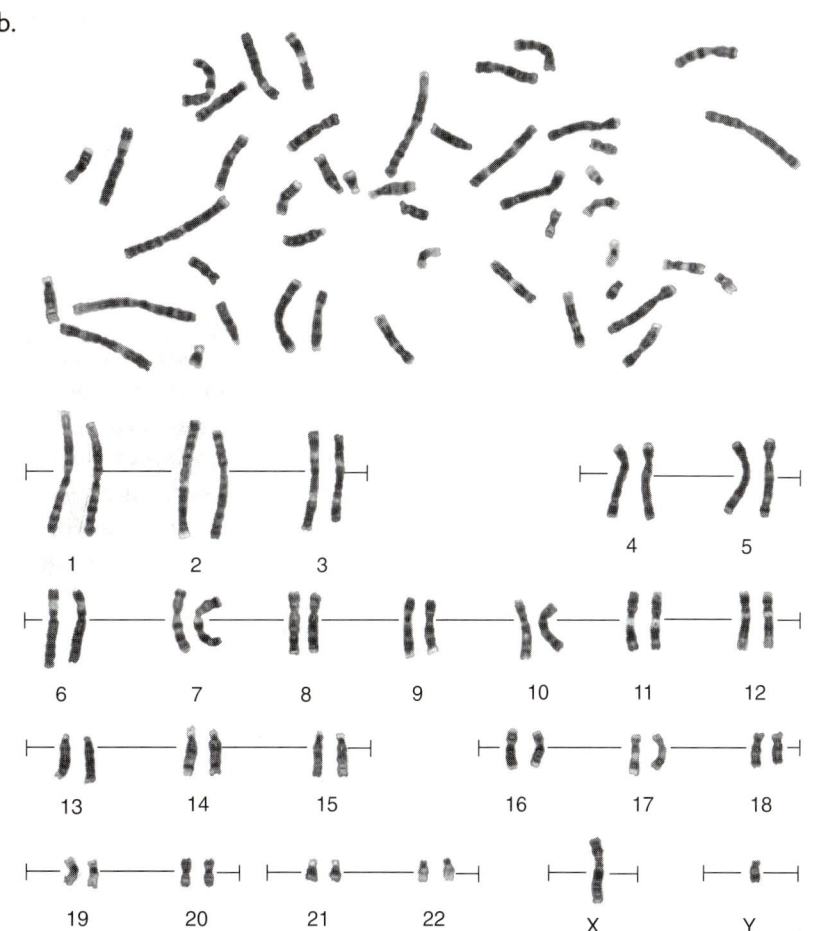

Figure 14.10

The thread-like strands are chromosomes (a). If you remove the chromosomes from the cell, you find that they exist in matching pairs (b). Scientists must treat the chromosomes with special chemicals and instruments to do this.

and each sperm in the male body develops with 23 chromosomes, *not* 23 pairs of chromosomes.

> **Stop and Discuss**
>
> 1. Why is it essential that sex cells form with only one set of chromosomes (23) and not 23 pairs of chromosomes as found in body cells?
> 2. In terms of diversity, what advantage might there be to having sex cells with only one set of chromosomes?
> 3. Given this background information on cells and chromosomes, would you revise your explanation about how traits are inherited? If so, how? If not, why not?

The Chromosome Theory of Inheritance

Scientists have been developing an explanation for the inheritance of traits for a long time. In this section you will read some of the results of their work. Even though you have explored traits for only a brief time, you probably will recognize parts of your explanations in the currently accepted scientific model. Look for your ideas as you read. You also should know that the model we present here is incomplete. You will learn only the basic features.

Scientists call the current model the **chromosome theory of inheritance.** It is so named because chromosomes contain pieces of information on their DNA called **genes,** which contain the information for inheritance.

According to the chromosome theory, each chromosome in a cell consists of many genes, with different genes containing information for the inheritance of different traits. So for example, as in Figure 14.11, one chromosome might have genes that are responsible for mid-digital hair, genes that are responsible for bony tori, genes that are responsible for freckles, and so on.

Figure 14.11

This is a sketch of a chromosome showing the locations of genes that determine particular traits. Geneticists (scientists who specialize in the study of genes and inheritance) study chromosomes and perform experiments to try to find the locations of genes on chromosomes. This example shows the way geneticists locate a gene on a chromosome.

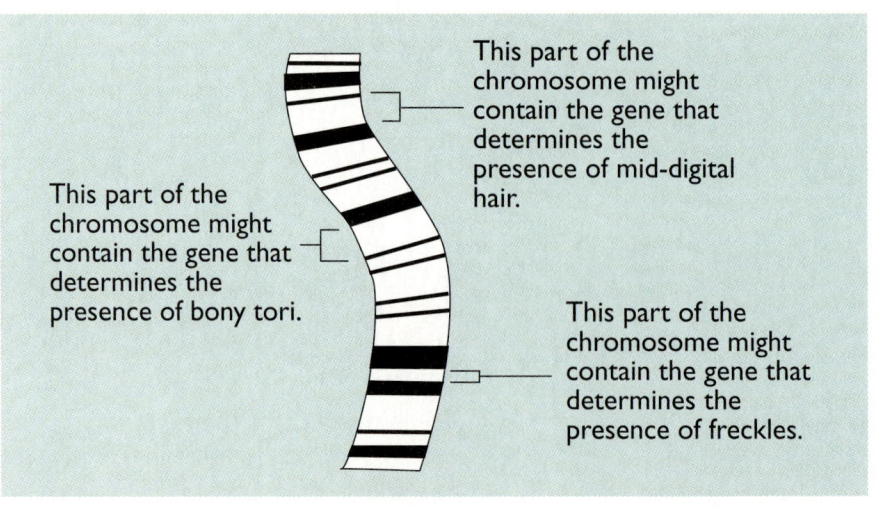

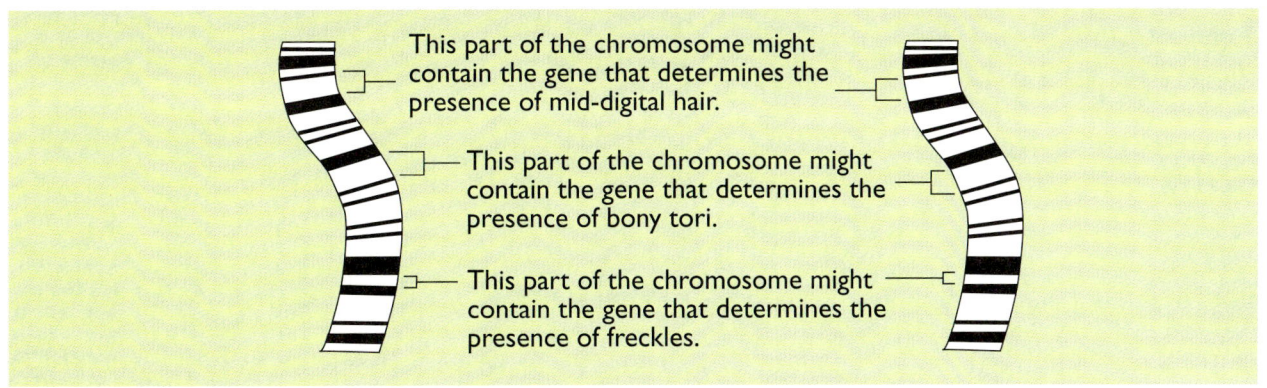

Figure 14.12 ▲

The chromosome from Figure 14.11 would exist as one of a pair, as we have drawn here. The genetic material from each chromosome together produces the observable trait.

Remember that chromosomes in body cells exist in pairs, so it is more accurate to draw pairs of chromosomes, as in Figure 14.12. The genes from each chromosome in the pair together produce observable traits.

The chromosome theory is easier to understand if we focus only on one trait at a time. To simplify your investigation of chromosomes in this reading, we will draw chromosomes as shown in Figure 14.13.

Study Figure 14.13. As with all scientific models, this model of chromosomes and genes is not an exact replica of the real thing. It does, however, allow us to explain observations and predict new occurrences.

Now study Figure 14.14, which represents a person's chromosomes containing the hereditary information for bony tori. There are two chromosomes, each carrying the information

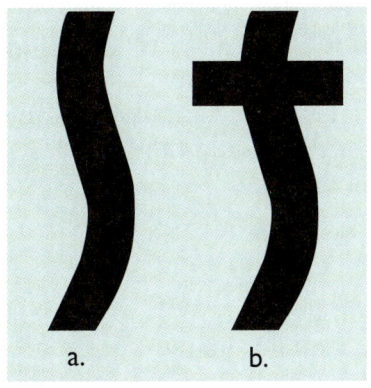

Figure 14.13 ▲

This is a model and only one way of representing chromosomes and where genes might be located on a chromosome. This is a sketch of a chromosome (a) and a chromosome indicating the location of a certain gene (b).

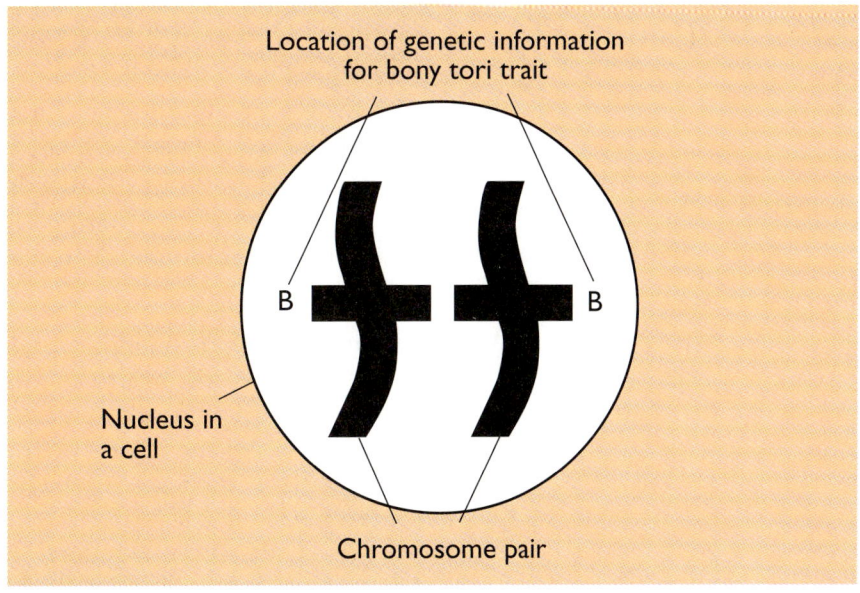

Figure 14.14

This is a model representation of a pair of chromosomes for the location of the gene that determines bony tori.

Explain ■ *Elaborate*

Figure 14.15

This is a model using only four chromosome pairs. Try to apply it to 23 chromosome pairs.

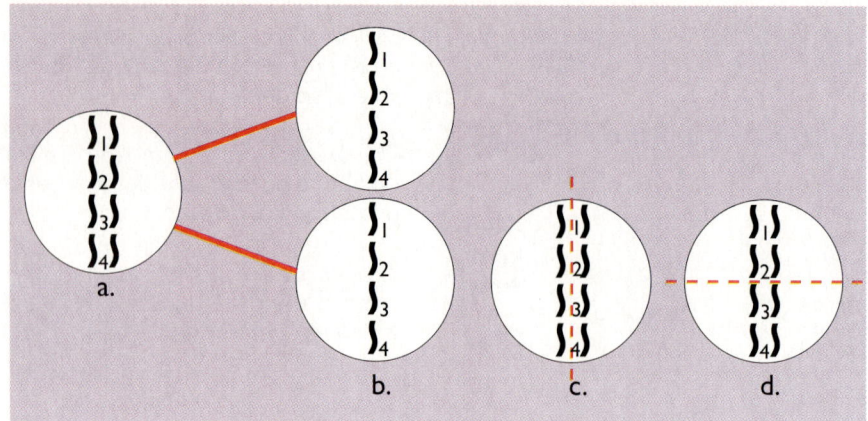

responsible for bony tori. The locations are labeled with a letter. We use letters to make it simpler to read and to make the model simpler to use. Scientists usually pick a letter that stands for the trait. In this case of bony tori, it makes sense to use the letter B for bony.

Notice in Figure 14.14 that the bony tori trait is the result of two Bs, each in the same relative location on each chromosome in the pair.

Stop and Discuss

4. Where do you suppose the pair of Bs depicted in Figure 14.14 comes from?

   Figure 14.15a shows chromosomes 1 through 4 in the nucleus of a cell before it divides into two sex cells. We did not draw all the chromosomes in this diagram. In a real human cell, all 23 pairs would be present. Figure 14.15b shows two of the sex cell nuclei that result.

5. How do you suppose the chromosomes in a cell divide to form a cell with only half of the chromosomes? Would they divide as in 14.15c or as in 14.15d? Justify your answers.

6. With this new information, how can you revise your own explanation or model about how traits are inherited?

Remember that scientific models explain many, if not all, observations. We still need to explain one observation that Marie made. For a given trait, children who resemble their parents usually have the trait of one or the other parent, but not both the parents' traits at the same time if the parents have different traits. For example, in Marie's case, her father has a cleft chin and her mother does not. Marie wondered why she had her father's trait for a cleft chin but not her mother's trait for a smooth chin. If her father gave her a gene for a cleft chin, and her mother gave her a gene for a smooth chin, she thought she should have a very small cleft chin, or something else that would be a combination of the two genes that she inherited from her parents.

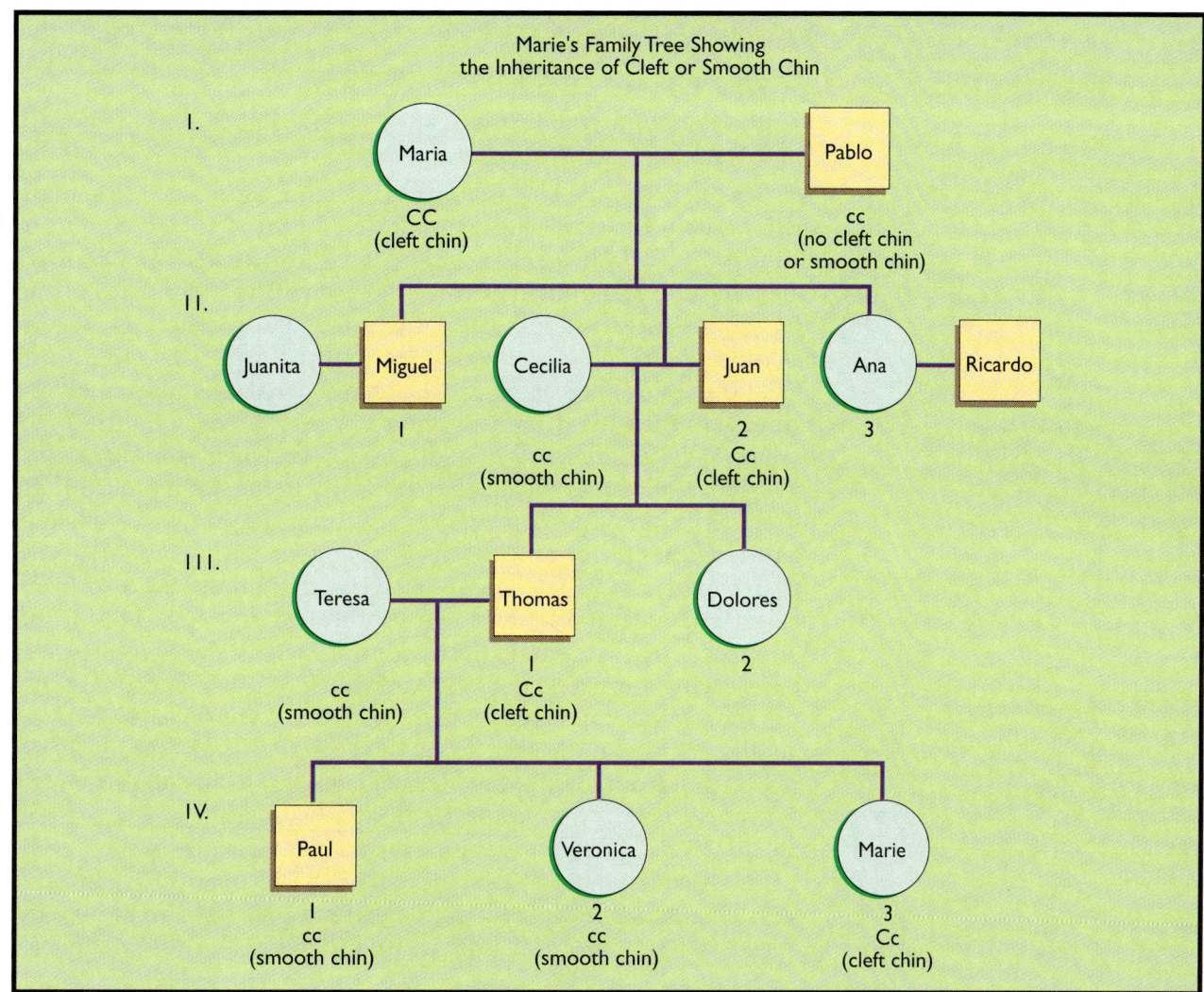

Figure 14.16

Study the inheritance of alleles for the shape of the chin in Marie and her ancestors.

The chromosome theory of inheritance explains this observation as follows: In humans, most traits such as eye color, weight, or hair color are determined by more than one gene. The trait Marie is concerned with, the shape of the chin, is determined by only one gene. But a gene can have a number of different forms, called **alleles** (al LEELZ). The gene involved in the shape of the chin has two alleles—one that produces a smooth chin and one that produces a cleft chin.

Like Marie, a person inherits one allele for chin shape from each parent. So each person has two alleles for chin shape. In this case Marie inherited an allele that produces a cleft chin from her father and an allele that produces a smooth chin from her mother. The question remains, why then does Marie have a cleft chin and not a smooth chin?

In the family tree in Figure 14.16, the allele for a cleft chin is shown as an upper-case C. The allele for a smooth chin is shown as a lower-case c. Remember the alleles are shown in pairs because a person has two—one from each parent.

Stop and Discuss

7. What do you notice about the presence of a smooth or cleft chin and the number of C or c alleles?

Scientists use the words "dominant" and "recessive" to explain the inheritance pattern shown in Marie's family. The cleft chin form of the trait is **dominant,** meaning it is "expressed" or "shown" whenever the allele for that trait is present. The other form of the trait, in this case a smooth chin, is expressed only when the person has inherited two alleles for that form of the trait. This other form of the trait is the **recessive** form, meaning that it is hidden or not expressed when the dominant allele is present. Study Figure 14.17 to help you understand the difference between dominant and recessive. Notice that the allele that results in the dominant form of the trait is marked by an upper-case C, and the recessive allele is marked with a lower-case c. This is how geneticists mark alleles. It doesn't matter in what order you write the letters, Cc or cC; the presence of a capital letter always indicates the expression of the dominant form of the trait. Study Figure 14.17. Paul, Veronica, and Marie each inherited one allele from their mother and one allele from their father for shape of chin. Paul and Veronica each received one allele for smooth chin from their mother and one allele for smooth chin from their father. Because neither had an allele for cleft chin, smooth chin was expressed. Because smooth chin is recessive, both alleles need to be present for the trait to be expressed.

To enhance your understanding of the idea of dominant and recessive traits, you will use the creatures you have undoubtedly noticed in your room. These creatures are known as "reggers."

Figure 14.17

Marie: This combination (a) gave Marie a cleft chin because the cleft chin allele is dominant—only one allele needs to be present for the trait to be expressed. Paul: This combination (b) gave Paul a smooth chin because smooth chin is recessive and two smooth chin alleles need to be present for the trait to be expressed. Veronica: This combination (c) also gave Veronica a smooth chin.

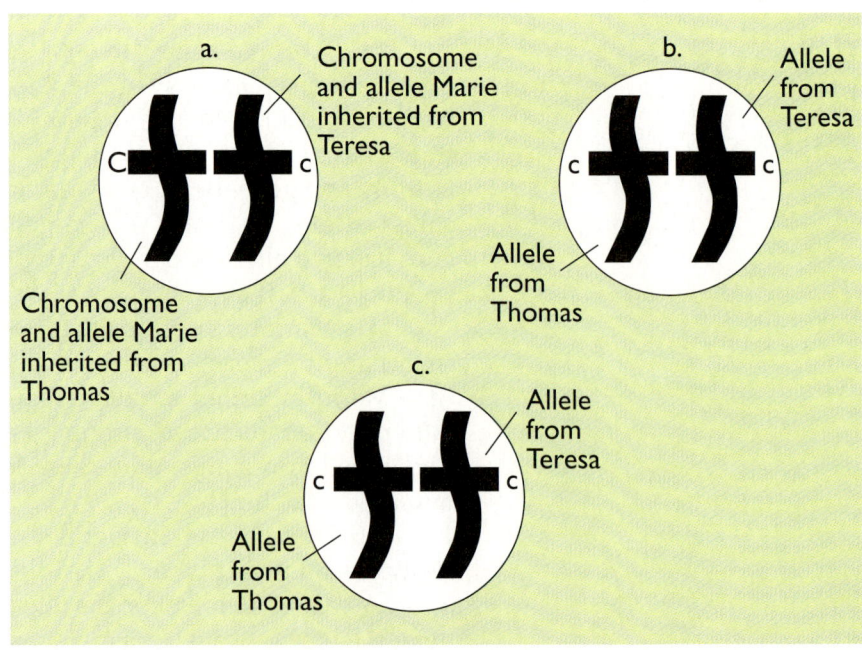

Figure 14.18

This is a magnified nucleus of a regger's cell showing genes for different traits. One gene is responsible for each trait. Each gene has two alleles.

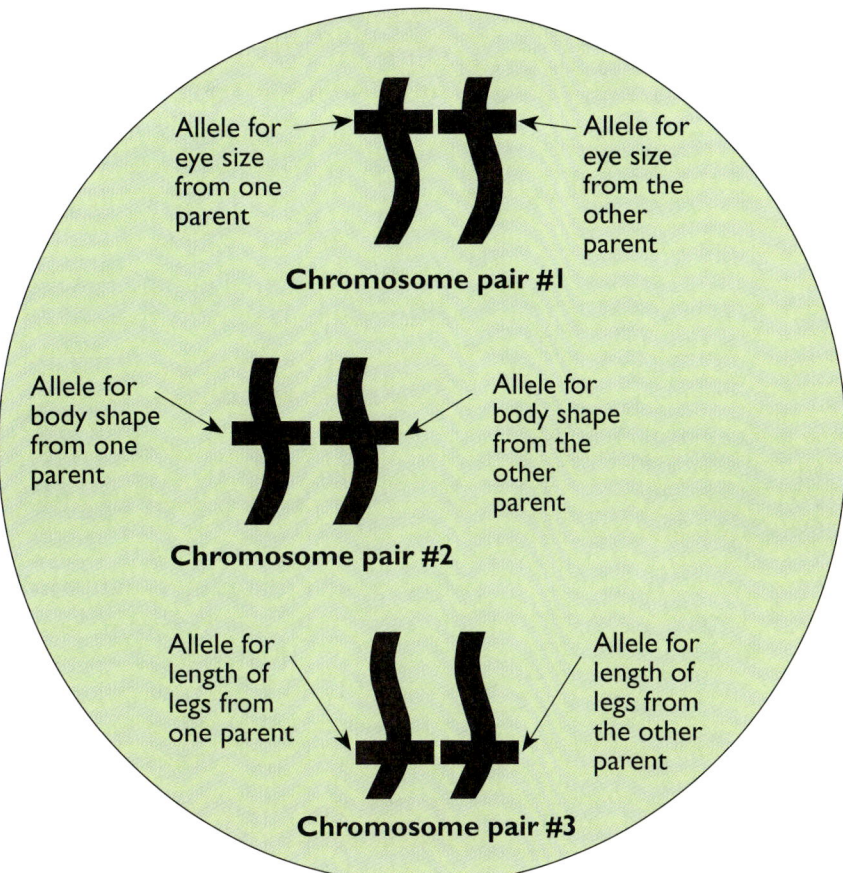

Imagine that reggers live in a society like yours. They go to school and to work. They have families. Some reggers are doctors, some are plumbers, some stay home with the children. They have their own complete society. Reggers have three traits that you will examine during this reading: eye size, length of legs, and body shape.

Reggers have six chromosomes (or three pairs of chromosomes) in their body cells. They inherit three chromosomes from each parent, which come together to make three pairs in their body cells. Each chromosome pair contains the gene for one of the three traits, as shown in Figure 14.18.

Ma regger has the big eye form of the eye size trait, the round body form of the body shape trait, and the long leg form of the leg length trait. Pa regger, on the other hand, expresses the traits with these forms: small eyes, oval body, and short legs. From their many genetic studies, regger scientists working at the famous *Regger Institute of Genetics* have learned that big eyes are dominant and small eyes are recessive. To simplify the drawings and this reading, we will refer to the allele for big eyes with an upper-case B. We will refer to the allele for small eyes with a lower-case b. Because the form of the trait for big eyes is dominant, a regger can have small eyes only if two alleles for small eyes are present. Only one allele for big eyes needs to be present for the regger to have big eyes. Two alleles for big eyes also results in the expression of big eyes.

Stop and Discuss

8. In your notebook draw the possible combinations of alleles for eye size that Ma could have, given her big eyes. Share and justify your ideas with the class.
9. Why can't you be *sure* of Ma's gene combination for eye size?

Pa also inherited one gene for eye size from each parent. Pa's eyes are small.

Stop and Discuss

10. What are the possible allele combinations that result in Pa's small eyes?

Let's try another trait: body shape. The same regger scientists learned that the round body form is dominant and the oval body form is recessive. Ma has a round body. Let's label the allele for a round body upper-case R for "round." Let's label the allele for an oval body lower-case r. For the body to be oval, the cells cannot contain any R genes, otherwise the R form of the trait would be expressed. Why? Because the round body form is dominant.

Stop and Discuss

11. In your notebooks draw Ma and Pa's allele combinations for the body shape each expresses. Share and justify your answers.

The last trait is length of legs. Regger scientists know that the short leg form is dominant and the long leg form is recessive. Let's label the short leg allele upper-case S for "short" and the long leg allele lower case s. Pa's body shows the dominant form of the trait. Ma's body shows the recessive form of the trait.

Stop and Discuss

12. Take a moment with your teacher to figure out Ma's and Pa's possible gene combinations for length of legs.
13. Help your teacher construct a data table to organize the information for the trait and allele combination that you have determined for all of Ma and Pa's traits.

Reggers have developed laws for their society. As in human society, there are certain laws that reggers must obey before they marry. One of the laws states that before marriage, each regger must visit a regger scientist to have a complete genetic workup. This means that the scientist removes some cells and determines, with the use of advanced technologies, what the regger's *exact* genetic combinations are for all of his or her traits.

Figure 14.19

Ma is Bb, Rr, and ss for regger traits. Pa is bb, rr, and Ss for regger traits.

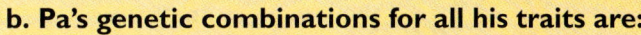
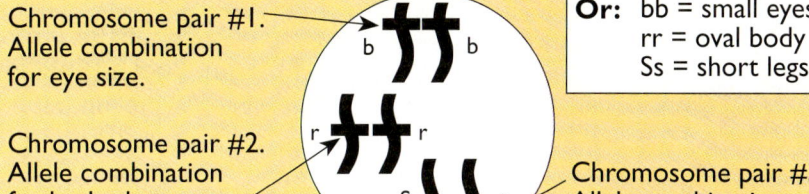

a. **Ma's genetic combinations for all her traits are:**

Chromosome pair #1. Allele combination for eye size.

Chromosome pair #2. Allele combination for body shape.

Chromosome pair #3. Allele combination for length of legs.

Or: Bb = big eyes
     Rr = fat body
     ss = long legs

b. **Pa's genetic combinations for all his traits are:**

Chromosome pair #1. Allele combination for eye size.

Chromosome pair #2. Allele combination for body shape.

Chromosome pair #3. Allele combination for length of legs.

Or: bb = small eyes
     rr = oval body
     Ss = short legs

Before Ma and Pa married, they visited a scientist for their genetic workups. Ma already knew that her genetic combination for her long legs was ss, but she did not know for sure whether her genetic combination for big eyes was BB or Bb. She also did not know whether her genetic combination for her round body was RR or Rr. Pa, on the other hand, knew two of his genetic combinations for sure. He knew that his combination for small eyes was bb and that his combination for an oval body was rr. He just wasn't sure if his genetic combination for short legs was SS or Ss. After their visits to the regger scientist, they knew for sure all of their genetic combinations. Figure 14.19 shows you what they found out.

Stop and Discuss

Use the chromosomes your teacher has just handed you to answer the following questions in your notebook. Your teacher then will call on you for your responses.

14. If Ma's sex cells must contain half of each pair of chromosomes from her body cells, how many different combinations could the body cell chromosomes sort into to make sex cells? What are the combinations?

15. If Pa's sex cells must contain half of each pair of chromosomes from his body cells, how many different combinations could his body cell chromosomes sort into to make sex cells? What are the combinations?

Notice that beside each regger, your teacher has placed some bowls. Ma has four bowls and Pa has two bowls. Each bowl represents one of the possible sex cells you just determined for Ma and Pa. In each of these bowls, your teacher has placed the chromosomal combinations that might be found in each sex cell, just as you determined in class. Your class will now conduct a short activity with these bowls.

Stop and Discuss

16. How many different types of young reggers could Ma and Pa have according to the results in your class?
17. Why did you choose numbers out of hats instead of choosing the sex cells directly as you may have wanted?

Notice that some of the young reggers looked like one or the other parent. Some looked like each other, as you might expect of brothers and sisters. The amazing thing is that with just three traits, three genes (each having two alleles), and two parents, there are eight possible types of offspring. This is the way that the chromosome theory accounts for similarities and differences among siblings and among children and their parents. The same explanation applies for humans, only each time two humans mate, they are not passing on just three genes. Geneticists estimate that there might be 100,000 genes in each human body cell. While Ma regger could make four types of eggs, and Pa regger could make two types of sperm, the potential number of different types of human sex cells is enormous when you consider that there are 23 pairs of chromosomes and 100,000 genes that can sort out during the process of making sex cells. The chance meeting of genes from one person's sex cells with new combinations of genes from someone else's sex cells could result in children with so many different gene combinations that it boggles the mind! And the likelihood of any two people who are not identical twins of being genetically identical is almost nonexistent.

Another observation that Marie made was that traits seem to skip generations. How can we have traits that neither parent has but that we can see in some of our ancestors?

Again the answer to this question concerns the idea of dominance and recessiveness. Let's go back to Ma and Pa regger, but we have to change a few things for this example. Let's say that Pa had big eyes, not small eyes; Ma stays the same. She too has big eyes, and she still carries the allele for small eyes. Her genetic combination is still Bb for eye size. The new Pa has big eyes, but his genetic combination for eye size is different from Ma's combination. His combination is BB.

Stop and Discuss

18. Using only the genes for eye size, draw the sex cells this Ma regger could form and the sex cells that this Pa regger could form. Share your ideas.

19. In your notebook determine all the possible combinations of Ma's eggs with Pa's sperm. Remember that we are concerned only with eye size for this explanation.

Suppose Ma and Pa have an offspring. They name her Uba, and she has big eyes. Also suppose that Uba has the genetic combination Bb, for big eyes. Uba grows up and falls in love with Dollo. Dollo has big eyes too, and the same gene combination that Uba has: Bb. They have a grand wedding, and shortly after, they have an offspring named Borg. Uh-oh. Borg has small eyes. Dollo is furious. He can't understand why Borg has small eyes when neither he nor Uba, nor any of the grandparents, have small eyes. Maybe their offspring got mixed up with someone else's offspring in the hospital nursery.

Stop and Discuss

20. How can you explain what happened with the size of Borg's eyes? Use the chromosomes for Dollo and Uba that your teacher hands out to determine the sex cells. Then determine the combination of Dollo's and Uba's sex cells that could show up in the offspring. Can you conclude that this offspring is really theirs?

Do you think Dollo and Uba were ever sure that this offspring was theirs? Well, thanks to the chromosome theory they were. To this day, Dollo and Uba are happily married and now have nine offspring, three of whom have small eyes.

Now that you've learned about the model that explains diversity, the chromosome theory of inheritance, how would you answer Marie's questions about the traits she inherited? Turn to the last paragraph of the connections section, Traits and Trees, review Marie's questions, and see how you can use your understanding of the chromosome theory of inheritance to answer her questions.

INVESTIGATION:
Too Tall or Too Short for Your Genes?

Are you female or male? Are you a thiourea taster or nontaster? Do you have a cleft chin or a smooth chin? You can be sure about your answers to these questions. Try another question: Are you short or tall? In this investigation you will elaborate on your understanding of genetics by looking at a trait that is not as easily defined as those you have studied already.

Explain ■ *Elaborate*

Working Environment

You will work individually and as a class. You will share a work space at a wall with your classmates. As you work find opportunities to practice the skill Praise others.

Materials

For the entire class:
- 1 wide roll of masking tape
- 15 metric tape measures or meter sticks

For each student:
- 1 white card, 3-by-5 in.
- 1 colored card, 3-by-5 in.
- 4 pieces of transparent tape

Procedure—Part A: Are You, or Aren't You?

1. As a class help your teacher construct a set of axes to make a giant graph on the wall. You will graph the number of females and males in your class.

   Your teacher will use your suggestions for labeling and scaling the axes.

2. Help plot the data on the graph by taking turns taping your white index card to the appropriate place on the graph.

   Apply tape to the back of the index card. Make sure that you do not tape your card on top of someone else's card. If you pay attention to what is on the vertical axis, as well as what is on the horizontal axis, you will avoid taping your card on top of someone else's.

3. Discuss the female and male graph according to your teacher's questions.

   Be sure to share your own ideas and questions as well.

4. After your teacher changes the trait on the horizontal axis to say "taster" and "nontaster," retrieve an index card and place it on the form of the trait that describes you.

   Use a new piece of tape if the used piece is not sticky anymore. Remember that you should pay attention to what is on the vertical axis to avoid piling your card on top of others' cards.

5. Discuss the taster and nontaster graph as directed by your teacher.

   Feel free to share your own observations and ideas as well.

6. After your teacher changes the trait on the horizontal axis to say "cleft chin" and "smooth chin," retrieve an index card and place it on the form of the trait that describes you.

   Use a new piece of tape if needed.

7. Discuss the cleft chin and smooth chin graph as directed by your teacher.

8. Leave the cleft chin and smooth chin graph intact.

Procedure—Part B: Back to Normal

1. With the aid of a partner, measure your height in centimeters.

   Notebook entry: Record your height.

2. Help your teacher construct the axes for a graph showing the distribution of height in your class.

3. Plot the data on the graph by taking turns taping your colored index card to the location on the graph that describes your height.

   Again pay attention to the scale on the vertical axis.

4. Discuss the height graph as directed by your teacher.

Wrap Up

Work on the following questions individually, recording your answers in your notebook. Prepare to share your answers in a class discussion.

1. How many different types of individuals are represented in the height graph?
2. Think of an explanation for why the number of types of individuals represented by the height graph is different from the number of types of individuals represented by the graphs in part A.
3. How would you define normal for traits such as height?
4. Where does "short" end and "tall" begin? Are you short or tall?

CONNECTIONS: Create a Tree

How well do you understand the idea of dominant and recessive traits? By creating a family tree, you can tell whether you understand how people inherit a dominant or recessive trait. Work with a partner to create a family tree in your notebook as follows.

1. Choose either Al, Ros, or Isaac as the subject of your tree.
2. Create four generations in the character's family. The character must have great-grandparents, grandparents, great aunts and uncles, parents, aunts and uncles, and siblings.
3. Draw the relatives on a tree using the symbols as described in the key to Marie's family tree in Traits and Trees.
4. Name the relatives.
5. Think of one trait you want the character to express. It does not have to be a trait you explored in the chapter. *Be creative.*
6. Assign that trait to the character and decide whether it is a dominant or a recessive trait.
7. Assign the trait to the character's relatives and ancestors to show how the character got the trait.
8. Write a paragraph describing the trait, whether it is dominant or recessive, and how the trait was passed on. Justify the inheritance pattern of the trait and present your tree to the class when your teacher directs you to do so.

Elaborate ■ *Evaluate*

CHAPTER 15

Genes and Society

Now that you know something about the chromosome theory of inheritance, you are ready to face some important issues and decisions related to the model. As scientists learn more about genes, they realize how powerful that knowledge is. Do you think there is power in knowledge? What kind of power can there be in knowing about genes? In this chapter you will investigate the power of the constantly growing knowledge of genetics, and you will face some issues that could affect future generations of humans.

INVESTIGATION: Designer Reggers

Have you ever heard the saying, "You can't fool Mother Nature"? What if you could fool Mother Nature? What if you could change genes and make something act or appear in a manner different from how it otherwise would act or appear? In this investigation you will have a chance to fool Mother Nature.

Working Environment

Work cooperatively in your team of two. Use the roles of Manager and Communicator. Push your desks together to create a work space or work side by side at a table. Practice the skill Praise others and remember to Disagree with the idea and not the person.

Materials

For each team of two students:
- Ma's and Pa's chromosomes
- Ma's and Pa's sex cell ovals
- 2 pairs of scissors
- school glue or glue stick
- several pieces of colored paper including orange, blue, pink, white, green, and red
- 2 pieces of large yellow construction paper
- 1 piece of large construction paper, any color
- your choice of colored markers

Procedure

1. Read the following scenario.

   You and your partner are reggers who are geneticists at the Regger Institute of Genetics. Recent technological advancements at the institute have made it possible for regger geneticists to change regger traits by manipulating and changing genes. The first case study was completed one month ago. The head geneticist managed, after years of grueling work, to help a male and a female regger produce an offspring with big *square* eyes. Because of these advancements, Reggerland now has a new law stating that in order for regger geneticists to change traits, they must have a very good reason for doing so. The head geneticist, for example, had a reason for producing a regger with square eyes. Scientific studies showed that in reggers square eyes filter out more of the sun's harmful rays than do the typical round eyes.

   As a regger geneticist, you not only have a license to change regger traits, but the Congress of Reggerland also has charged you to do so. You and your partner will now collaborate in changing a regger trait. You have been specifically assigned to work with Ma and Pa regger.

2. Discuss the three traits reggers have. In your discussion address the following questions:

   a. Is there a benefit to having one particular form of any one of these traits? If so what might that benefit be?

   Look back to the reading A Model That Explains Diversity in Chapter 14 to review the three regger traits.

b. If you could change any of these regger traits, how would reggers then benefit?

Notebook entry: Record your answers to these questions.

3. Choose one trait that you and your fellow geneticist will change in an offspring of Ma and Pa regger. Adhere to the following criteria:
   - *You must produce only one new form of a regger trait.*
   - *You must adhere to regger law and justify your reason for producing the new form of the trait.*
   - *You must prepare a presentation for the Congress of Reggerland when you are finished. (Coincidentally, the Congress of Reggerland calls such a presentation a "Wrap Up.")*

   Notebook entry: Record the trait you chose to change, the change you will make, and your justification for the change.

4. Conduct a brainstorming session to decide how you will change Ma's and Pa's current genetic makeup to produce an offspring that has the new desired trait.

   As you think of strategies, you should consider the following questions:
   - *Where will you make the changes: in the body cells or in the sex cells?*
   - *If you need to introduce a new gene, which form of the trait will be dominant and which form of the trait will be recessive?*

   Notebook entry: Record all the strategies you can think of.

5. From the list you generated in step 4, choose one strategy to try.

6. Using your team's strategy, change the genes to produce a young regger with the trait that you want.

   Cut, add, remove, or paste genes or alleles as necessary. Use yellow paper for new alleles or to indicate where and how you made changes. Rather than cut any of your teacher's sets of chromosomes or sex cell ovals, make your own copies of those you will need on the appropriate colored paper, cut them out, and use these copies for your changes.

7. Simulate matings by combining Ma's and Pa's sex cells.

8. If all of the possible combinations of the matings do not produce the trait you want, choose a new strategy from the list you developed in step 4 and start the process again.

   You might have ideas for completely new strategies you did not think of in step 4. If you need help, ask the Communicator to gather information for you. Also be sure to share your ideas with Communicators from other teams who might come to you for help.

9. Draw a picture of what Ma's and Pa's new young regger will look like, complete with the new trait that you produced.

   Attach the parts of the chromosomes and sex cell ovals or other materials on which you made changes in your notebook to save them.

Wrap Up

Prepare and conduct a presentation for the Congress of Reggerland. Include the following information in your presentation.

1. Discuss your reason for changing Ma's and Pa's genetic makeup to produce a new trait. Tell the Congress why the new trait is desirable and how the new trait improves reggers.

2. Using your materials, show and describe to Congress the process you went through to make sure that the offspring would show the desired trait. Be ready to explain why you made the choices you did.

3. Tell the Congress of Reggerland about the working policies at the *Regger Institute of Genetics.* This means that you must praise others and disagree with the idea and not the person. Tell Congress how you and your fellow scientists put these policies into practice as you worked together.

4. You also will listen to the presentations of other teams of geneticists. As you listen remember to praise the ideas you think are praiseworthy, ask questions, and offer helpful comments or advice.

READING:
Genetic Engineering

In the previous investigation, you had a chance to simulate technology that might be available in the not-too-distant future. Right now there are scientists called **genetic engineers** who can change the genetic makeup of an organism to produce a new or different trait. This process is known as **genetic engineering.** Genetic engineers often use one-celled organisms such as bacteria. (Bacteria is the plural form of the word bacterium.)

Compared to other organisms, bacteria are genetically simple organisms. They have one circular chromosome that is not contained in a nucleus. Genetic engineers use chemicals to open this chromosome and insert new genes. Figure 15.1 shows a model of how genetic engineers can change the genes on a chromosome ring in a bacterium.

When a bacterium has the new genes, it may express the new trait or traits associated with these genes. One of the most successful examples of genetically engineered bacteria is a kind that produces human insulin. What makes it possible for bacteria to successfully express a human trait is that all organisms on earth use

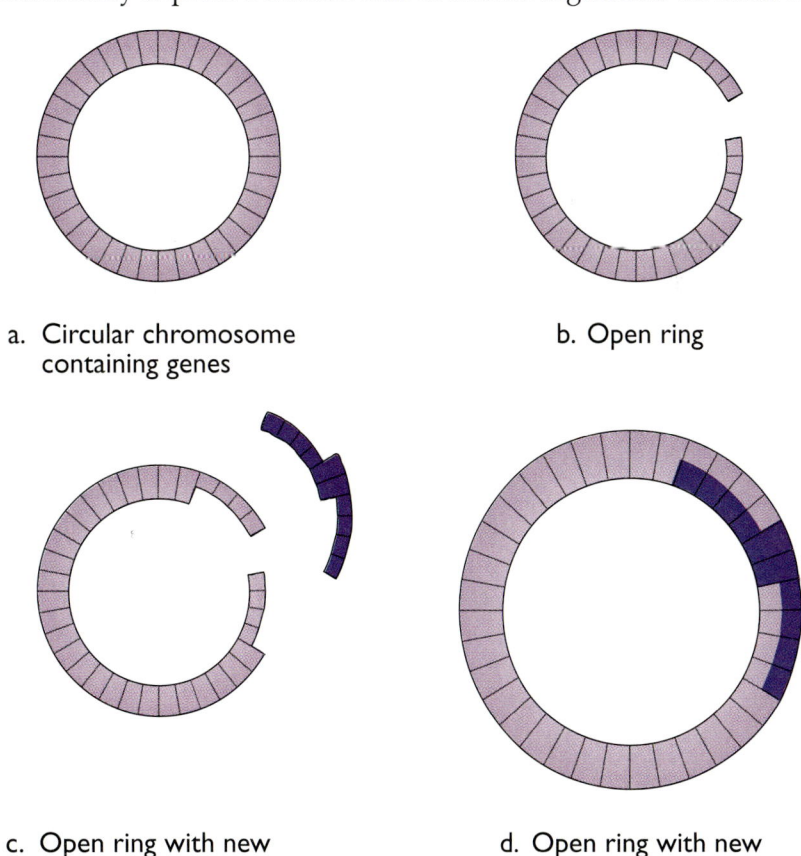

a. Circular chromosome containing genes

b. Open ring

c. Open ring with new genes coming in

d. Open ring with new genes

Figure 15.1

Genetic engineers can use chemicals to open the chromosome in a bacterium to insert new genes.

Explain

DNA as an information particle. Furthermore the DNA of all organisms is the same because all organisms are related.

Insulin is a substance that your body produces to help control the amount of sugar in your blood. Some people do not produce enough insulin to control the amount of sugar in their blood. These people have a disease called diabetes. If the blood-sugar level in a diabetic is not controlled, it causes reactions that range from thirst to disorientation (not knowing where you are) to coma, and sometimes to death.

For years people with diabetes have been able to give themselves regular injections of insulin to keep their blood-sugar levels under control. In the past scientists extracted this insulin from an organ in cows, sheep, and pigs. Genetic engineering, however, has changed that. Now that genetic engineers can use bacteria to produce *human insulin*, diabetics can use a substance that is a closer match to their own bodies. Diabetics are better able to tolerate human insulin than cow, sheep, or pig insulin. When using human insulin, diabetics don't need to use as much insulin as they did when they were using insulin from other animals. This makes the treatment of diabetes less expensive. In the future genetic engineers hope they will be able to change the genes of diabetic people so they can produce enough of their own insulin.

Genetic engineering technology is widely used in agriculture, too. For example, genetic engineers can alter the genes of a certain harmless bacterium to resist the formation of frost around them until the temperature drops to ⁻5° C. Normally, frost forms at about 0° C. Genetic engineers can then spray strawberry plants with this bacteria, and the bacteria help the plants resist freezing until temperatures drop below ⁻5° C. This increases the chance that the plants will survive a late frost. As a result, farmers can produce more strawberries.

Scientists also have learned how to replace certain genes in cows so that they produce smaller dairy cows that produce *more* milk than large dairy cows. Farmers then can use less feed for smaller cows yet produce more milk. Farmers are thus able to increase their profit in two ways: buying less food and selling more milk.

Advances such as these encourage genetic engineers to continue their research and experimentation. Consider the following advances in genetic research that are currently underway.

Determining Gender

Recent news reports claim that a group of genetic engineers in London, England were able to genetically change female mice to male mice. They operated on newly fertilized female eggs and injected them with the gene that they think is responsible for determining sex. When the mice were born, they indeed appeared to be males. As they grew, they acted like male mice in many respects.

Cloning

People sometimes talk about clones wishfully. "Gee, if only I had a clone, then my clone could do all of the things that I hate to do!" What is a clone? A **clone** is an exact genetic replica of an organism, gene, or cell. Identical twins are clones. In order to make a clone of you, genetic engineers would have to know how to make or duplicate all of your genes in a laboratory. They then would have to know how to put all of those genes together in the right order to produce your clone. Even if your clone might be genetically identical to you, however, it might not act or even exactly resemble you if it lived in an environment different from yours.

So far genetic engineers have not been able to clone people. But they have observed natural clones and actually have cloned some kinds of amphibians. Bacteria make clones of themselves to produce new bacteria. Cloning also occurs in some plants. For example, by planting the eyes of a potato, we can produce another identical potato. Also, strawberries send out runners a short distance from the original plant and produce a new plant replica. The roots of rose bushes, lilac bushes, and aspen trees grow out underground to sprout replicas of themselves close by. All of these are examples of cloning in nature.

DNA Fingerprinting

Geneticists now can use DNA to help solve crimes or mysteries. How? Remember that genes are made of DNA. Also remember that every organism's DNA is made of the same material components. But each individual is unique because the arrangement of these DNA components differs from person to person. Scientists can treat and study a sample of DNA, obtained from skin cells, blood cells, or body cells of other tissues and fluids, and then match it to the person to whom it belongs. Because the arrangement of material in each individual's DNA is as unique as a fingerprint, the use of DNA to solve mysteries is called DNA fingerprinting.

Here's an example of a mystery solved by DNA fingerprinting:

In 1990 seven people in Florida who had AIDS accused their dentist, who also had AIDS, of infecting them with HIV (an AIDS-causing virus). This was the subject of much debate and controversy. If the dentist, who has since died, did indeed infect his patients, this could mean that patients of health-care providers with AIDS are at risk for contracting HIV. So scientists set out to determine whether the patients were infected by that dentist or by someone else.

It is important for you to realize that cases like these are very rare. It also is important for you to realize that all health care providers, including dentists, must follow specific safety procedures (such as wearing disposable latex gloves) to protect their patients and themselves from unknowingly passing on or contracting viruses.

Scientists and disease specialists know many things about viruses, including that viruses have their own genetic material. Typically a virus infects a person by injecting its own genetic material into the cells of the host. The host then incorporates the virus's genetic material into its own DNA.

Scientists also are gaining much information about the AIDS virus including the following: (1) there are different strains of HIV; (2) each strain of HIV has its own unique genetic combination; (3) when one person infects another person with HIV, the parts of the DNA in their cells that have incorporated the genetic material from the HIV will look more similar than the infected DNA from the cells of two people who were each infected by someone else with different strains of HIV.

Disease specialists obtained a sample of a cell infected with the HIV from the dentist in question. They extracted the infected portion of the DNA, treated it, and determined its DNA arrangement. These scientists then obtained samples of the HIV-infected cells from the seven patients in question. The scientists extracted the DNA from the infected cells and determined the DNA arrangements in the HIV of the seven people.

When they compared the arrangements of the HIV DNA from the dentist to the HIV DNA from the seven patients, they found the results that are shown in the following table in Figure 15.2.

Patient	Dentist's HIV DNA
A	Match
B	Match
C	No Match
D	Match
E	No Match
F	Match
G	Match

Figure 15.2

This table shows the results of the tests that compared the dentist's HIV DNA to the seven patients claiming that he had infected them.

Source: Stroh, M. (1992, May 23) DNA Sowes Aids Epidemiological Whodunit *Science News*, P. 341

Stop and Discuss

1. What conclusions could the scientists draw? Could they now conclude that the dentist infected five patients and did not infect the other two patients?

Scientists do not believe that DNA fingerprinting technology is refined enough yet to be completely accurate without a doubt. This technique, however, does have potential to become an extremely accurate means of identification.

DNA Dog Tags

The United States military has decided to use DNA for identifying soldiers. "Dog tags" are the traditional metal identification tags soldiers wear around their necks. These metal pendants are stamped with a soldier's name, social security number, blood type, and religion. The military now will require that a sample of blood be taken from each soldier to extract the DNA from a cell. They will keep a picture of that DNA on file for every soldier. These "DNA dog tags" will be a record of the unique material structures in DNA for each soldier. If a soldier dies and the remains are unidentifiable, experts could match a sample of DNA from the body to the sample on file to help positively identify the soldier. This would reduce the incidence of soldiers who are unidentifiable after death or those listed as missing in action.

Stop and Discuss

2. Consider all of the examples of genetic engineering in this reading. In what ways might these technologies increase the power of human society?
3. In what ways could this increased power be negative?
4. Read the following section, The Future of Genetics—The Future Is Now. Then discuss these questions:
   a. What power is there in mapping the human genome?
   b. What responsibility comes with that power?

The Future of Genetics—The Future Is Now

In 1990 the United States Congress passed legislation to fund the **Human Genome Project.** A **genome** is all of the genetic information on all 23 chromosomes (each human nucleus contains two copies of the genome for a total of 46 chromosomes). One of the purposes of the Human Genome Project is for geneticists to locate the chromosomal positions of all the roughly 100,000 genes in the human genome. On which chromosome pair and at which spot on the chromosome pair are the genes for bony tori? mid-digital hair? freckles? As they discover the locations of individual genes on chromosome pairs, geneticists are creating a map to show other scientists and interested individuals where they can find any specific gene. Geneticists then will try to determine the exact makeup of each human gene, including the unique arrangement of the DNA material components.

What is the point of knowing the chromosomal locations of all the human genes? The more we understand about human diversity, the more advanced we will become in areas such as diagnosing heritable diseases. This might bring us closer to curing such diseases as cystic fibrosis, diabetes, and even cancer. It also might help us understand more about the theory of human evolution. It also could bring us closer to the type of genetic engineering in which human traits could be altered.

Scientists think that the Human Genome Project will take at least 15 years to complete. After the project is complete, the knowledge of the location of the genes on chromosomes could enhance our power to improve human life. But imagine: what if the results of this project created new opportunities for discrimination? For example, what if you wanted a job, but no one would hire you because employers asked for a copy of your genetic profile, and they did not like the kind of genes you had? They then might discriminate against you and deny you a job. They might, for instance, dislike the fact that you had genes that indicated you had a high risk of developing heart disease at an early age.

What if there were a law that required you to present a copy of your genetic profile before you could get a marriage license, just as Ma and Pa Regger had to? What if, at that time, the person who distributes marriage licenses was required to report you to certain authorities if you carried a gene that might produce a serious disorder in your children? Would the authorities let you get married? If they let you get married, would they let you have children? Could they stop you? Should they stop you?

The Human Genome Project is underway right now. It is already a part of the present and will be an important part of your future. Look for news on the Human Genome Project in newspapers and magazines.

CONNECTIONS:
Society and Genetic Engineering

Do you foresee a time when genetic engineering will be unlimited—a time when genetic engineers would have the technology to make any changes with genes? Before you answer this question, first consider the following. Then be ready to share your ideas and participate in a class discussion.

1. Genetic engineers currently work under a variety of constraints. Consider the following partial list of constraints and explain why each of these is considered a constraint.

   a. money
   b. ignorance
   c. social standards
   d. personal beliefs and values
   e. technology
   f. fear
   g. regulations and laws
   h. limits to the ability of genes to determine traits—for example, how the environment plays a role in the expression of traits

2. Which of the above constraints, if any, do you think might not exist in the future? Be ready to justify your response.

3. Which of the above constraints might become more important?

4. If genetic engineers could change the sex of humans before they were born, which sex do you think most people would want their babies to be? Why?

5. Suppose that in the future genetic engineers could change children's genes. For which traits do you think parents would most request that their children's genes be changed?

6. Referring to question 5, if scientists could change genes, would this automatically result in a change in the trait? Explain your answer.

INVESTIGATION:
Science Fiction?

Imagine a time when genetic engineering technology is much further advanced than it is presently. Imagine what a day would be like, what you'd see as you walked down the street, and how life would be different from what it is now. Would life be better? more fun? more exciting? Would life be awful? dull? like a nightmare? What would constrain the technology? Would there be laws? Would law officers be able to enforce those laws? Keep imagining. That's what this investigation is all about.

Materials

For each team of four students:
- any supplies you choose from those that your teacher provides

Working Environment

Work individually in Part A. In Part B meet with your partner and combine with another team of two to form a cooperative team of four. You will need a large work space. As you work in Part B, practice the unit skill.

Elaborate

Procedure—Part A: Writing

Write a story that fits the following criteria:

- Your story must begin with these lines: "It is the year 2092. It has been 20 years since genetic engineering technology took an enormous step forward. Let me tell you what it is like now."
- The story must be about what you would see and what the world would be like in 2092 if we had been using advanced genetic engineering technology for 20 years.
- You must choose at least three of the constraints from the connections Society and Genetic Engineering. You must assume that these constraints are still important in 2092. You must include a description of the constraints and describe how they are affecting genetic engineering in 2092.
- You must incorporate some of your ideas about what is right and wrong with genetic engineering.
- You must be someone living at that time in that society.
- You must write your story at an elementary reading level.

Notebook entry: Record your story in your notebook.

Procedure—Part B: Designing

1. Design a book that will hold all your Team Members' short stories by using a process that includes creating a table for criteria, constraints, and decisions.

   Compare each other's flow charts for the design process that you created in Unit 2. Consider the following as you create your table:

   - *What features do books typically have?*
   - *What makes a book appealing?*
   - *What do the contents of a book usually look like?*

2. Present your book as your teacher directs.

Wrap Up

1. Present your book to the rest of the class and read your stories to your classmates.

2. After the presentation your team should present your finished version of a paragraph that begins with the following unfinished sentence:

   "To summarize my cooperative learning experience, _____."

   (Try to include what cooperative learning is to you, what you have accomplished through cooperative learning, and how you feel about cooperative learning.)

SIDELIGHT

Science Nonfiction

So you wrote a science fiction story. Many authors before you have done the same. Some authors specialize in science fiction. Some authors who wrote science fiction never realized that the things they imagined might come true. Consider the following examples:

- In the 1950s Ray Bradbury's book *Fahrenheit 451* told of large video screens, high-speed automobiles, interactive TV sets, and government burning books it did not approve of.

- In the 1800s Jules Verne wrote books titled *Twenty Thousand Leagues Under the Sea* and *Journey to the Moon* in which he told of adventures in underwater ships and giant artillery shells that could fly to the moon.

- In the 1940s George Orwell's book *1984* told of a society in which the government secretly placed listening devices in buildings. Using the devices, the government could listen in on what people were doing or saying.

Do any of these fictional things ring true? If you read these books and science fiction works written by other authors, you might find more examples of science fiction turned nonfiction. What did you write about in your science fiction story? Were you careful? You never know. You could have just written science nonfiction!

Elaborate

HOW TO
Construct a Data Table

#1

In the investigation Star Tracers you recorded data in a data table and then used that data to analyze your results. If you had written the information haphazardly anywhere in your notebook, you might have had a much harder time analyzing which attempts were successful and which were unsuccessful. Because you used a data table to organize your data, however, all you had to do was fill in the blanks and count up the total number of successes. You also had all the information you needed to answer the wrap-up questions and participate in the class discussion.

Scientists frequently use data tables to organize the information they collect in their investigations. That way they can be sure they have recorded everything, and they can easily locate the information they need. Organizing data in a data table makes the final job of analyzing data easier, so that the scientists or the

students who use the data tables can learn from their investigations. After all, learning is the purpose of conducting investigations.

What if constructing the data table had been your responsibility in the investigation Star Tracers? Would you have known how to organize such a table? In this How To, you will learn, step by step, how to construct a data table so that you can make your own data tables in future investigations.

The following steps will help you construct a data table for almost any investigation. The examples with each step will show you how Marie, Isaac, Al, and Ros organized their data tables for the investigation Star Tracers. You can follow their examples and the general steps in organizing your data table for the next investigation, Threading the Needle.

1. Read all of the steps in the investigation.

   If you read all the steps before you begin, you can plan ahead, because the steps in the investigation tell you what data you need to collect as you go along.

2. Decide on the problem you are trying to solve or the question(s) you are trying to answer through this investigation.

Figure H1.1

What do you think of Ros's and Al's sample list of data they might need to collect?

Ask yourself the question, as Marie and Isaac did, "What am I supposed to find out from this investigation?"

3. List the kinds of data you must collect.

   Decide what you will need to keep track of during the investigation. Will you need to keep track of the time, the date, measurements, people's names, different objects? Be sure to list everything you might possibly need to know during and after the investigation.

4. Review your list from step 3 in light of the problem or question you identified in step 2, and decide which items on your list will help you solve the problem or answer the question(s) you identified in step 2.

   Remember that it is better to have too much information in a data table than too little. If you record too little information, you stand the chance of having to repeat parts or all of the investigation to obtain what you left out or forgot. If you have too much information, you might not need it at all, but it only cost you a few minutes of extra work. A good scientist keeps complete records.

5. Draw and label the columns of your data table.

   Decide how to organize your revised list from step 4 into the columns of your data table. Each column must have a heading that relates to

Figure H1.2

Ros and Al are editing their list.

one item from your list. Think about the order of the columns. What information do you already know? (For example, you already know the names of your Team Members.) What data will you collect first, second, and third? Organize the columns of your data table into a logical sequence with clear, easy-to-read headings.

Our Star Tracers Data Table			
Person doing it	How long before they finished or gave up	What happened during the try	Was it a success? Yes or No
Ros	2 minutes	She gave up	No
Ros	2 minutes, 10 seconds	She went out of the track 3 times	No
Ros	3 minutes, 30 seconds	She took forever	Yes
Ros	3 minutes	She almost went out of the track	Yes
Ros	3 minutes, 3 seconds	Same thing	Yes
Al	10 seconds	He quit	No
Al	8 seconds	Hand wouldn't move	No
Al	15 seconds	He quit	No
Al	5 seconds	He broke the lead	No
Al	11 seconds	He crumbled the paper	No

How to Construct a Data Table

Also, it is wise to use an entire sheet of paper for each data table you draw. You need to be sure you have enough room for all your data.

6. **Give your data table a title.**

   The title of the data table should relate to the investigation or experiment for which you are collecting the data. Ros might title her data table "Our Star Tracers Data Table."

7. **Complete as much of the data table as you can before you begin the investigation.**

   If you can fill in any parts of the data table ahead of time, you will save time and effort during the investigation. For the investigation Star Tracers, Ros and Al could fill in the first column of their data table before they began. (See Figure H1.3.) Which columns can you complete in advance in your data table for the investigation Threading the Needle?

Figure H1.3

Complete in advance as much of the data table as you can.

Our Star Tracers Data Table			
Person doing it	How long before they finished or gave up	What happened during the try	Was it a success? Yes or No
Ros			
Ros			
Ros			
Ros			
Ros			
Al			
Al			
Al			
Al			
Al			

Our Star Tracers Data Table			
Person doing it	How long before they finished or gave up	What happened during the try	Was it a success? Yes or No
Ros	2 minutes	She gave up	No
Ros	2 minutes, 10 seconds	She went out of the track 3 times	No
Ros	3 minutes, 30 seconds	She took forever	Yes
Ros	3 minutes	She almost went out of the track	Yes
Ros	3 minutes, 3 seconds	Same thing	Yes
Al	10 seconds	He quit	No
Al	8 seconds	Hand wouldn't move	No
Al	15 seconds	He quit	No
Al	5 seconds	He broke the lead	No
Al	11 seconds	He crumbled the paper	No

Figure H1.4

This shows the data table that Ros and Al completed.

8. Record the data in the appropriate boxes as you complete the investigation.

As you complete other investigations in this program, you will have to organize your own data tables. In each case, try to follow the steps described here. Use the example from the investigation Star Tracers to help you visualize what a new data table might look like.

HOW TO
Construct a Graph

#2

A good graph is worth a lot of words! Graphs are useful tools because they provide us with "pictures" of information that often communicate more clearly than words. We organize graphs in a particular way so we can easily read the information the graphs contain. Graphs also help us find patterns in the data we have collected.

There are certain steps you should follow in constructing a graph, just as there were certain steps you followed in constructing your data tables. The following steps will help you construct your graphs for the class data in the investigation Threading the Needle.

1. Review the data you have recorded.

   To construct this sample graph, you should review the class data for the investigation Threading the Needle. Those data should be recorded in the class data table that your teacher has posted on the board or on an overhead transparency.

2. Draw the horizontal axis and the vertical axis for the graph.

   *Graphs have two lines, one that runs back and forth (horizontally) across the page and one that runs up and down (vertically). These lines have special names. We call the line that runs back and forth the **horizontal axis** and the line that runs up and down the **vertical axis**. The point where these two lines, or **axes**, meet is the place where the graph begins. (The term **axes** is the plural for the term **axis**.)*

   It helps if you draw your axes on graph paper, because graph paper provides evenly spaced lines and squares. Use a ruler to help you draw straight lines. Draw the vertical axis close to the left-hand edge and the horizontal axis close to the bottom edge of the graph paper. Make the lines longer than you think you need. This will make your graph easy to read.

3. Label each axis by using the headings in the data table.

   *The labels on each axis tell about the **variables** you investigated in the experiment. (A variable is anything that can change in an experiment or investigation.) You label each axis for a different variable so that you can see the relationship between the two variables. The variables that you want to relate on a graph of your class data are the number of students and how many successes the students achieved during the investigation Threading the Needle.*

   Label the vertical axis "Number of Students" and the horizontal axis "Number of Successes," as Isaac has done.

4. Set up the number scales on each axis.

   *Both axes of a graph often have a sequence of numbers called a **number scale**. You read the numbers on the horizontal axis from left to right and those on the vertical axis from bottom to top. In your*

graph for Threading the Needle, the numbers on the horizontal axis are the number of times students were successful in threading the eye bolt, while the numbers on the vertical axis are how many students had a certain number of successes. Write a zero just outside the left-hand corner of your graph, where the vertical and horizontal axes meet.

To decide what numbers to use in your number scales, look at your class data table. Determine the number of successes that **most** of the students in the class had. Count the number of students who had that number of successes, and write that number next to a line at the top of the vertical axis. Also from the class data table, determine the greatest number of successes that any **one student** had, and write that number in a space at the far right on the horizontal axis. Number as many of the lines on the vertical axis and the spaces on the horizontal axis as you think you need to plot your data accurately. (Compare the number scales on the graphs in Figure H2.4. Notice that Isaac did not number every line and space.) You should leave equal spaces between whole numbers. The objective is to make a big, clear picture of the class data from the investigation Threading the Needle. Try not to squeeze the graph down toward the bottom or over to the far left. Fill the entire page with your graph.

The number scale on one axis does not have to be exactly *the same as the number scale on the other axis. The difference between the numbers next to each other on an axis must be the same, though. Look at the following outlines of graphs that Al, Rosalind, Marie, and Isaac drew. Do you think all of their number scales are okay? Why or why not?*

5. Plot the data on your graph by doing the following:

- locate the first number on the horizontal axis,

    The first number on the horizontal axis is one. That stands for one success in threading the eye bolt. Look at your class data table. How many students were successful one time?

- trace your finger up the column above the label to the place on the vertical axis that shows where that piece of data fits,

    Let's say that three students were successful in threading the eye bolt one time.

- draw a horizontal line at the correct height to make the top of the bar,

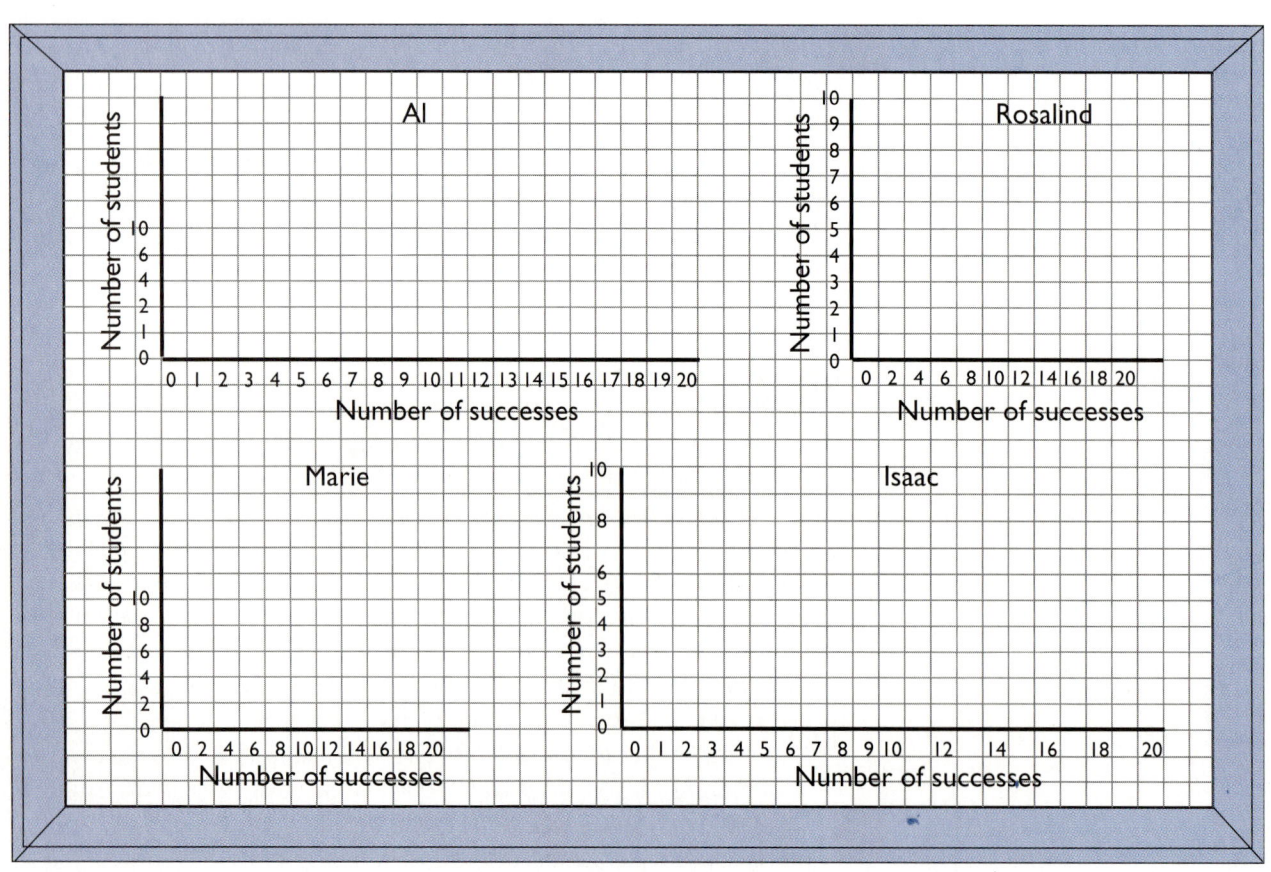

In this case, draw a line across the first space (#1) at the third line (#3) on the vertical axis.

- color in the bar from that line down to the horizontal axis, and
- repeat the parts of step 5 for all the pieces of data in the data table.

Next count the number of students who were successful at threading the eye bolt two times and draw a bar that represents those students. Then count the number of students who were successful at threading the eye bolt three times and draw a bar that represents those students. Continue this way until you have drawn bars for all the data in your class data table. Note that some numbers on the horizontal axis might not have a bar at all. There would be no bar if no one had that particular number of successes.

6. Finally give your graph a title.

   Another important feature of a graph is its title. The title should tell the reader something about the purpose of the graph. You might title

this graph something simple like "Class Graph of Data," or you could be more elaborate and call it "Our Class Success Story in Threading an Eye Bolt."

When you are finished, you will have a beautiful bar graph of what happened in your class in the investigation Threading the Needle. To see more clearly the pattern that the bars in the graph make, connect the tops of the bars with a continuous, smooth line. Your line doesn't have to dip and bulge dramatically with the bars.

The objective is to round out the bars to make a nice, smooth curve. What is the general shape of the curve that connects the tops of the bars?

Whenever you need to construct a bar graph in future investigations, refer to the steps in this How To. Even though the example is specific to the investigation Threading the Needle, the general steps should help you draw any bar graph.

HOW TO
Have a Brainstorming Session

#3

Having a brainstorming session can be a lot of fun. During a brainstorming session, you can voice any idea that comes to your brain, no matter how crazy it might seem. *Any* idea that you think might provide a solution is one you can propose. Sometimes the ideas that seem really far-fetched at first can lead to other ideas that really work! The main purpose of a brainstorming session is to create a storm in your brain so that you become really creative.

MARIE: I wonder where the word "brainstorming" came from.

ROS: The teacher said it had something to do with "divergent thinking."

ISAAC: According to the dictionary, to diverge means "to go or extend in different directions from a common point, to branch

out; to differ, as in opinion or manner; to depart from a set course or norm."

(Isaac is reading from a dictionary.)

AL: Good! That means I can diverge and express my own ideas, even if they are different from the rest of yours.]

As Ros indicated, the process of brainstorming involves divergent thinking, a term that comes from the root word "diverge." Isaac defined the term for you, and Al interpreted it correctly. During a brainstorming session, you have permission to be as different in your thinking from the others in the group as you like. Your goal is to come up with as many different ideas as possible. After you finish your brainstorming session, you can return to reality and use the ideas you thought of to solve a problem or to answer a question.

Before you begin your brainstorming session, decide how you will record your ideas. Remember that *no editing is allowed* during the brainstorming session. The editing comes later. When you are ready to begin your brainstorming session, follow these guidelines:

1. State any idea about the topic that comes to your mind.
2. Record everyone's ideas. Don't judge whether the ideas are good or bad. Write them all down.
3. Keep thinking of ideas for as long as you can or until your teacher tells you the time is up.
4. If you can't think of a new idea, try to add something to an idea that is already on the list. Do not change the idea that is on the list; just add a new twist or a new way of looking at that idea.
5. If you are working in a group, take turns. Be sure that each person has a turn to suggest ideas. Remember that some people might need a little more time to think before suggesting something.

After you finish your brainstorming session, look at your list and decide which ideas might be better than others for solving the problem or answering the question. This is when you can "edit" your ideas. You should have a great list from which to choose!

HOW TO
Conduct a Research Project

#4

Your task in the investigation Enabling the Disabled is to learn enough about one handicap so that you can redesign an environment to make it more accessible to people who have that handicap. One way you can do that is to find out all you can about the handicap or disability that you select. To find out about that particular handicap, you will do some research. There are two keys to conducting good research: (1) choose a topic that interests you, and (2) get organized! If you go about your research in an organized manner, then it will be much easier to put your information together to complete your task. This How To has some tips that will help you conduct your research.

You might be wondering what research is, anyway. How is conducting research different from completing other assignments? According to the *American Heritage Dictionary*, to research a topic, such as a handicap, means to study that topic thoroughly. Therefore conducting research is different from just reading an article in a magazine or looking up something in an encyclopedia. Also, when you conduct research, you do more than just read. You gather as much information from as many sources as you can about the topic you are researching. Then you put the information together in an organized way that will help you accomplish your task.

Because all of the Team Members will find out something about the handicap your team selects, you should divide up the responsibilities. The Tracker should make sure that each Team Member is responsible for a specific part of the research and that the teammates are not duplicating one another's efforts. Be sure to check in with one another often, so that you know you are on the right track and that others know what information you are finding. That way, too, you can find out if you need to change your plan before you waste too much time.

Tips for Conducting Research

Part A—Choosing Your Topic

Tip #1: List several topics (in this case, handicaps) that can help you accomplish your task.

Your team probably has a head start on this part because you already conducted a brainstorming session. The purpose of your brainstorming session was to decide on a handicap that you want to accommodate in the environment you chose. If your team has not conducted a brainstorming session yet, then continue with Part A. If your team has already chosen a handicap, then go to Part B.

To get started, you should list at least five possible handicaps that you could accommodate in the environment you will redesign. Then rank order those handicaps and circle your first and second choices. (You should always have at least two choices, because you might not be able to find enough information about your first choice.)

Stop now and conduct your brainstorming session. Make a list of five or more possible handicaps that you could study. You might write your list on BLM HT4.1, Planning for My Research on a Handicapping Condition.

Part B—Getting Organized

Tip #2: Think about your topic before you read anything.

You might ask yourselves these questions about the handicap you have chosen:

- What do we already know about this handicapping condition?
- What sources of information can we name that might have some information about the handicap?
- What do we *need* to know about the handicap?
- Is there anyone at school or in our neighborhoods who has this handicap?

If you organize your thoughts first, then it will be easier to locate information and organize the information you find.

You might organize your research by writing each of those questions, in addition to others you choose, at the top of one sheet of paper. Then as you find information that answers one of those questions, you can write the information on that sheet of paper. That way you can be organized before you even start your research.

As you read and talk to people, stay open to new ideas. You might think of new questions about the handicap that you would like to answer. You can start a new page for each question that you find interesting.

Stop now and write some things you already know about the handicap your team chose and some of the questions you would like to answer. You might write what you already know and your questions on BLM HT4.1, Planning for My Research on a Handicapping Condition.

Part C—Finding Information

Tip #3: Use more than one source for information.

Sources of information you might want to consider include the following:

- a person who has the handicap you chose;
- a hospital or medical supply company;
- a school for the visually or hearing impaired, a hospital, or a nursing facility;
- your school library, a community library, a university library, or a medical library; and
- yourself, if it is possible for you to assume the handicap.

This section will discuss each of these sources and how you might access them. You might want to list your sources on BLM HT4.2, Gathering Information for My Research.

A person who has the handicap you want to research. If you know of someone who has the handicap you want to research, you should set up an interview. Even if one of the team members knows this person, you should still organize a formal interview. You might interview the person directly and/or interview a member of the handicapped person's family. Before you arrange for an interview, review the following points:

- Call or personally contact the person and ask if he or she would be willing to talk with you.

  In your initial call, be sure to state your name and explain that you are conducting a research project for your science class. Explain that you want to find out how you could make an environment more accessible to a handicapped person.

  Ask the person how much time he or she would be able to spend with you and arrange for a specific time and place to meet.

  Once you arrange for an interview, be sure you write down the date, time, and place.

  Try not to be late for your interview.

- Write a list of questions that you plan to ask the person.

  Be sure the questions are specific to the handicap and to the environment you are redesigning.

  You might briefly share your list of questions when you arrange for your interview so the person can be thinking about answers.

- Ask direct questions such as, "Could you use a desk such as this one?" and show the person a photograph or a drawing of a desk from the environment you are redesigning or provide a description of the environment. Then ask, "If so, how would you use it? If not, how would you change it?"
- Be sure to show respect for the person you are interviewing. Do not treat him or her any differently from anyone else you might talk with. Remember, this person has important information to share with you.
- Thank the person for taking the time to talk with you. Offer to share your team's report if the person is interested.

A hospital or medical supply company that has relevant equipment. You might contact a medical or hospital supply company that has special equipment for a person with the handicap you are studying. When you contact such a company, you should do the following:

- State your name and the reason for your call;
- Ask if the company carries equipment that a person with the specific handicap might use;
- Ask about the store hours and when someone might be available to help you find out about specific equipment, such as a prosthesis (artificial limb) or a hearing aid;
- Set up an appointment with someone from the company;
- Arrange for transportation ahead of time to be sure you arrive on time for your appointment;
- Bring tape measures or meter sticks to take measurements of the equipment that might be essential for the handicap you are researching. For example, you might need to know the height and width of a typical wheelchair, walker, or crutches; and
- Be sure to thank the person who helped you.

A school for the visually or hearing impaired, a hospital, or a nursing facility. If the handicap you chose involves visual or hearing impairment, then you might want to visit such a school and talk with staff and students. Staff at local hospitals or nursing facilities also might be able to provide you with information or access to someone who has the particular handicap you are researching. Be sure you follow the same procedures outlined previously for conducting interviews and making appointments. Always let the contact person know who you are and why you are calling. Always be on time for any appointments you make, and thank everyone who helped you.

Your school library, a community library, a university library, or a medical library. Libraries contain a lot of information that might help you with your research project. Not only do libraries contain a lot of printed material, but they also have special sources of information, such as card catalogs, on-line (computerized) catalogs and databases, the *Reader's Guide to Periodical Literature*, encyclopedias, medical dictionaries and medical reference materials, audiovisual materials, and pamphlet files. If you have not used these types of resources before, ask for help from your media specialist, librarian, teacher, or a friend who has used them.

Before you use reference materials in a library, you need to know enough about your topic to identify a few *key words*. Key words are important words that are related to the topic you want to study. Sometimes you can find key words by reading an article in an encyclopedia or in a magazine about the specific handicap you chose. You also can come up with key words by interviewing someone or a family member of someone with the handicap. For example, if you chose the handicap of cerebral palsy, you might first look up the words "cerebral palsy" in an encyclopedia or in a dictionary. There you might find words such as "birth defect," "lack of coordination," and "muscle weakness." The words inside the quotation marks are all possible key words because they tell you something about the cause and effects of cerebral palsy.

You will need to have some key words in mind before you can effectively use the card catalog, an on-line database, or the *Reader's Guide to Periodical Literature*. You might list the key words you will use on BLM HT4.2, Gathering Information for My Research.

Yourself, if it is possible for you to assume the handicap. You should assume the handicap only within the environment you plan to redesign so that your direct experience will help your team complete its task. Depending on what handicap you chose, you might use a wheelchair, walker, or crutches; wear a blindfold; tie your hands behind your back; navigate on only one foot; or write with the opposite hand. Decide on a reasonable length of time to assume this handicap, and be sure you have taken safety precautions so that you do not injure yourself or someone else in the process. Keep track of problems you encounter, how you solve those problems, and how you feel with this handicap.

Stop now and list the sources you will use to get started. Then list a few key words that will help you find out more about the handicap your team has chosen. You can add to your list of key words as you do your research. Refer to BLM HT4.2 for help in organizing your information.

Part D—Organizing your information.

Tip #4: Reread your task and the questions you want to answer. Decide which information you have collected relates directly to accomplishing your team's task.

When you have collected enough information, your team needs to put together all the information and then decide which information helps your team accomplish its task. First have each

Team Member organize his or her information from most important to least important in terms of accomplishing the team task (redesigning the environment to accommodate the handicap you have been researching). Second ask each Team Member to read or describe the three most important pieces of information he or she found. Decide how you will keep a record of each Team Member's information. You might use BLM HT4.3, Organizing My Information, as a guide.

Continue sharing information until you have enough to begin work on redesigning the environment to accommodate the handicap you chose. Team Members can share additional information as you complete the investigation.

Glossary

adhesive forces: The attractive forces between unlike particles. In this book adhesive forces refers to the forces between particles of water and the particles that make up a paper boat.

afterimage: An afterimage occurs when a spot exists in your vision after you have seen a bright flash of light. Afterimage refers to the desensitization of the retina to certain colors so that you see only the complementary colors for a brief period of time.

alchemist: A person who practices alchemy.

alchemy: The branch of ancient science concerned with changing one material into another using an elusive material known as "philosopher's stone."

alcohol consumption: Alcohol consumption is the act of consuming or drinking alcohol.

alleles (al LEELZ): These are the different forms of a gene or trait.

atoms: This refers to the small particles that make up materials. The philosopher Democritus named these small particles atoms.

average: If you made a list of every person's height, added the heights together, and divided by the total number of people on your list, this would give you the average or the mean height for students in your class.

axes: The term axes is the plural for the term **axis**.

axis: An axis is a horizontal or vertical line in a graph on which we write numbers or labels. Most graphs have both a horizontal and a vertical axis, and the place where these two lines meet is where the graph begins.

blind spot: In the eyeball the portion of the retina that contains no rods or cones is called the blind spot. You do not see images that fall on the blind spot.

blood alcohol content (BAC): BAC is a measure of how much alcohol is in a person's blood. This is measured as a percentage.

body cells: These are cells other than sex cells that make up parts of the body such as skin, muscles, bone, teeth, and blood. They contain 46 chromosomes which exist as pairs.

chromosome theory of inheritance: This is the current scientific model that says chromosomes contain pieces of information on their DNA called genes and that these genes contain the information for inheritance.

chromosomes (KROH moh sohmz)**:** These are long threadlike strands of material found in the nucleus of a cell. They are made of DNA and contain the body's genetic information. All organisms on earth use DNA as an information particle.

clone: This refers to an exact genetic replica of an organism, gene, or cell.

cohesive forces: The attractive forces between like particles are called the cohesive forces. In this book cohesive forces refers to the attractive forces between water particles.

complementary: In this book the term complementary refers to the color that results from removing another color from the spectrum of light. For example, if you remove green from white light, then the rest of the colors blend to form the **complementary** color of green, which is red. This works for all colors. If you remove orange from white light, what remains is the complementary color of orange—blue. The complementary color of black is white.

comprehension: This term means "understanding."

cones: Cones are specialized cells in the retina of the eye that help you see colors.

continuous motion rate: This refers to the speed at which you must present a series of pictures in order to perceive continuous motion.

constraints: Human factors and other things that affect the **criteria** a designer sets for a product are called constraints. Constraints are limits a designer might encounter when trying to fulfill the criteria.

control variables: This refers to the steps you undertake in order to keep all variables in an experiment constant except for the one that you want to test.

criteria: In this book criteria refers to technology and design. Criteria is the plural form, **criterion** is the singular. Criteria are the goals that designers set for what a product will do, look like, act like, or be like. A product then is judged by whether or not it fulfills all the criteria.

D/s ratio: This is the value derived from the calculation of dividing the distance a person was from a screen of lines by the space between the lines. This value represents how far a person can stand from a screen to see a clear picture.

data: The data are the measurements or information that you collect during an investigation.

data table: A data table is a chart that helps a person keep track of observations or when making observations.

diabetes: This is a disease in which a person does not produce enough insulin to control blood sugar levels.

divergent: Divergent means branching out and having a different opinion.

diversity: The term diversity describes a variety in individual abilities. The word means "differences."

DNA or deoxyribonucleic (dee OK sih RY boh noo KLEE ik) **acid:** These are particles that make up chromosomes. Components of DNA make up genes.

DNA dog tags: This is a process the military uses to identify soldiers through the use of DNA.

DNA finger printing: This is the process by which experts use DNA from body fluids or tissues taken at the scene of a crime to identify an alleged perpetrator.

dominant: This refers to the form of a trait that is "expressed" or "shown" whenever one allele for that trait is present.

driving under the influence (DUI): A person driving under the influence is driving while intoxicated with alcohol or other drugs.

echo-location: The ability to use sounds, instead of vision, to locate objects.

electrons: Electrons are tiny particles in atoms.

ergonomics (ER goh nah miks)**:** This is the branch of science that studies how people interact with the products, facilities, equipment, environments, and procedures they use at work and in their homes. It is the science of human factors.

family tree: A family tree is a method of recording data about a family for many generations.

flicker-fusion frequency: Refers to how fast to present a series of flashes in order to achieve the appearances of **continuous motion**.

fovea: This is the area of the retina that contains the most rods and cones. You can see images more clearly if the image falls on the fovea.

function: In this book function refers to what a product can do. Function is a criterion.

genes: These are pieces of information on DNA found in chromosomes that contain the information for inheritance.

genetic engineering: This refers to the process scientists use to change the genetic makeup of an organism to produce a new or different trait.

genetic engineers: These are scientists who specialize in genetic engineering.

genome: This refers to all of the genetic information on all 23 chromosomes. Each human nucleus contains two copies of the genome for a total of 46 chromosomes.

graph: A visual representation of the results of your data.

hardness: The property of a material that is measured by how firm it is.

horizontal axis: The line in a graph that runs across the page from left to right.

human factors: Human factors are the differences in individual human beings that must be considered in setting standards, such as sizes, in order to ensure a proper "fit."

Human Genome Project: This is a government funded project set up to try to locate the chromosomal positions of all the roughly 100,000 genes in the human genome and to create a map to show these locations.

if–then statements: A logical statement that first states what a model is and then the results you would expect from an experiment.

inherit: This means that traits are passed down from generation to generation.

insulin: This substance, which is produced in the pancreas, regulates a person's blood sugar.

limit: The term limit means boundary or something that you can't go beyond.

majority: This term means "the most" of something. A majority of the population means more than 50 percent of the individuals in that population.

material: This refers to the makeup of a product or thing.

minority: This term means "the least" of something. A minority of the population means less than 50 percent of the individuals in that population.

normal curve: Normal curves, also known as bell curves, show the normal diversity in various limits for a group of organisms.

NTSC standards: The National Television Systems Committee (NTSC) is the organization that sets the standards we currently use to produce TV pictures in the United States and Japan.

nucleus (NOO klee us)**:** This is a part of a cell that contains the chromosomes.

number scale: A sequence of numbers plotted along a line is called a number scale. The number scale can be vertical or horizontal. The numbers usually are read from left to right on the horizontal axis and from the bottom to the top on the vertical axis.

operational definition: An operational definition is a standard definition of how you measure something.

particle model (or particle theory): This model presents the idea that all materials are composed of particles. The model is an estimate of what one of these particles might look like.

perception distance: The perception distance is the distance you travel during your perception time. This is the first phase in the stopping process.

perception time: The perception time is the time that elapses between hearing and perceiving.

peripheral vision: This is the ability to see around you without turning your head and while looking straight ahead.

persistence of vision: Persistence of vision is when the human visual system retains an image it sees for a very short time after that image is no longer on the retina.

phenomena: Phenomena means happenings or occurrences. It is the plural of **phenomenon.**

philosopher: A philosopher is a person who thinks about things and develops explanations for them.

pitch: In this book this term refers to the shape and twist of a certain propeller.

pixels: The term pixels refers to small sections that appear as dots on a grid that covers the TV or computer screen. Pixels come in three colors: red, green, and blue.

property: A property of a material can be the look, feel, function, or other unique characteristic of a material.

range: A range defines the inner and outer limits of a characteristic such as peripheral vision. Ranges exist in individuals as well as within populations.

reaction distance: This is the distance you travel during your reaction time, before you begin to stop. This is the second phase of the stopping process.

reaction time: The time it takes for one to react after perceiving an event.

recessive: This refers to the form of a trait that is hidden or is not expressed when the dominant allele is present. Both recessive forms must be present for the trait to be expressed.

research: Research means to study a particular topic thoroughly. When you research you gather information about the topic from as many sources as you can.

retina: The retina is the innermost layer of the eyeball. The retina contains the rods and cones.

rods: Rods are specialized cells in the retina of the eye that help you see in the dark.

scientific model: A scientific estimate that is a representation of something people cannot directly or easily observe accurately is called a scientific model.

sex cells: These are cells other than body cells. They are the eggs in females and the sperm in males. These cells only have 23 chromosomes, which do not exist as pairs.

skidding distance: During the skidding time, you travel a distance known as the skidding distance. The skidding distance is the third and final phase of the stopping process.

skidding time: The skidding time is the time that passes between when you first try to stop and when you finally come to a complete stop.

surface tension: This refers to the special way that particles of water are cohesively attracted to each other.

technology: Technology refers to designing products that fit people and help them solve problems.

testability: This refers to the component of models that classify them as science or non-science. If a model is testable and does not include elements of magic, superstition, faith, or other components that are impossible to test, the model is classified as science.

theory: An explanation for a set of observations or a group of phenomena.

total stopping distance: This refers to the distance you travel during the total stopping time. You can derive the total stopping distance by adding the perception distance, the reaction distance, and the skidding distance.

total stopping time: This is the entire time it takes you to come to a complete stop. You derive the total stopping time by adding the perception time, the reaction time, and the skidding time.

traits: This refers to the characteristics of living things.

translucence: This is the property of a material that is measured by how much light passes through it.

variable: A variable is anything that can change in an experiment or investigation. A variable can affect the results of your experiment.

vertical axis: The line in a graph that goes from the bottom of the page to the top of the page is called the vertical axis.

viscosity: This refers to the property of a material that is measured by how easily something flows or pours.

(Board Members continued from p. ii.)

Tracy Posnanski, *University of Wisconsin-Milwaukee, Milwaukee, Wisconsin*
Douglas Reid, *Southridge Middle School, Fontana, California*
Rochelle Rubin, *Instructional Materials Center, Waterford, Michigan*
Charlotte Schartz, *Kingman Elementary School, Kingman, Kansas*
M. Gail Shroyer, *Kansas State University, Manhattan, Kansas*
Elayne Shulman, *Classroom Consortia Media, Metuchen, New York*
Barbara Spector, *University of South Florida, Tampa, Florida*
John Staver, *Kansas State University, Manhattan, Kansas*
John Swaim, *University of Northern Colorado, Greeley, Colorado*
Robert Tinker, *Technical Education Research Centers, Cambridge, Massachusetts*
David Trowbridge, *University of Washington, Seattle, Washington*

Project Advisors and Consultants

William D. Gillan, *IBM, Boca Raton, Florida* (Corporate Advisor for Design Study)
Martin Guttmann, *IBM, Boca Raton, Florida* (Corporate Advisor for Design Study)
Ann Haley-Oliphant, *Mainville, Ohio* (Contributing Author)
Norris Harms, *Arvada, Colorado* (Evaluation)
A. W. Harton, *IBM, Atlanta, Georgia* (Corporate Advisor for Design Study)
James McClurg, *University of Wyoming, Laramie, Wyoming* (Curriculum Development)
Ann Primm, *Knoxville, Tennessee* (Contributing Author)
James R. Robinson, *Boulder, Colorado* (History)
M. Gail Shroyer, *Kansas State University, Manhattan, Kansas* (Implementation)
Dave Somers, *Colorado Springs, Colorado* (Editor)
Terry G. Switzer, *Fort Collins, Colorado* (Contributing Author)
Luise Woelflein, *Washington, DC* (Contributing Author)

Field-Test Sites
Primary Site Centers and Affiliated Schools

California
Almeria Middle School, Fontana, California, 1990–91
Southridge Middle School, Fontana, California, 1990–92
Coordinated by Herbert Brunkhorst (Site Coordinator) and Carol Cyr (Graduate Assistant 1990–91) and Cynthia Peterson (Graduate Assistant, 1991–92) based at California State University, San Bernardino, California.

Colorado
Carmel Middle School, Colorado Springs, Colorado, 1990–92
Challenger Middle School, Colorado Springs, Colorado, 1990–92
Colegio Los Nogales, Bogota, Colombia, South America, 1991–92
The Colorado Springs School, Colorado Springs, Colorado, 1990–92
Desert School, Rock Springs, Wyoming, 1991–92
Eagleview Middle School, Colorado Springs, Colorado, 1990–91
East Junior High School, Rock Springs, Wyoming, 1991–92
Gorman Middle School, Colorado Springs, Colorado, 1990–92
Panorama Middle School, Colorado Springs, Colorado, 1990–92
Smiley Middle School, Denver, Colorado, 1991–92
Timberview Middle School, Colorado Springs, Colorado, 1990–91
White Mountain Junior High School, Rock Springs, Wyoming, 1991–92
Coordinated by BSCS staff based in Colorado Springs, Colorado.

Florida
Clearwater Comprehensive School, Clearwater, Florida, 1990–91
Harllee Middle School, Bradenton, Florida, 1991–92
Lincoln Middle School, Palmetto, Florida, 1991–92
16th Street Middle School, St. Petersburg, Florida, 1990–92
Southside Fundamental School, St. Petersburg, Florida, 1990–92
W. D. Sugg Middle School, Bradenton, Florida, 1990–92
Coordinated by Barbara Spector (Site Coordinator) and Merton Glass (Graduate Assistant) based at University of South Florida, Tampa, Florida.

Kansas
Chapman Middle School, Chapman, Kansas, 1990–92
Dawes Junior High School, Lincoln, Nebraska, 1991–92
East Junior High School, Lincoln, Nebraska, 1991–92
Fort Riley Middle School, Fort Riley, Kansas, 1990–92
Kingman Middle School, Kingman, Kansas, 1990–92
Murdock Elementary School, Kingman, Kansas, 1990–92
Norwich High School, Kingman, Kansas, 1990–92
Norwich Junior High School, Kingman, Kansas 1990–92
Pound Junior High School, Lincoln, Nebraska, 1991–92
Coordinated by John Staver (Site Coordinator), Randall Backe (Graduate Assistant, 1989–91), and Ronald Krestan (Graduate Assistant, 1991–92) based at Kansas State University, Manhattan, Kansas.

New York
Roy W. Brown Middle School, Bergenfield, New Jersey, 1991–92
Longwood Junior and Senior High School, Middle Island, New York, 1990–91
Longwood Middle School, Middle Island, New York, 1990–91
Mount Sinai Middle School, Mount Sinai, New York, 1991–92
Shoreham-Wading River Middle School, Shoreham, New York, 1990–91
Southampton Intermediate School, Southampton, New York, 1991–92
Tremont School, Mount Desert, Maine, 1990–91
Coordinated by Thomas Liao (Site Coordinator), Rita Patel-Eng (Graduate Assistant, 1989–91), and Cynthia Anderson (Graduate Assistant, 1991–92) based at State University of New York, Stony Brook, New York.

Ohio
Dater Junior High, Cincinnati, Ohio, 1990–91
McCord Middle School, Worthington, Ohio, 1991–92
Perry Middle School, Worthington, Ohio, 1991–92
Pleasant Run Middle School, Cincinnati, Ohio, 1990–92
Coordinated by Glenn Markle (Site Coordinator) and Cynthia Geer (Graduate Assistant) based at University of Cincinnati, Cincinnati, Ohio.

Secondary Site Centers and Affiliated Schools

Arizona
Lee Kornegay Junior High School, Miami, Arizona, 1991–92
Tso Ho Tso Middle School, Fort Defiance, Arizona, 1991–92
Williams Middle School, Williams, Arizona, 1991–92
Coordinated by Diane Ebert-May (Site Coordinator) and Alison Graber (Graduate Assistant) based at Northern Arizona University, Flagstaff, Arizona.

California
Hollenbeck Middle School, Los Angeles, California, 1990–91
Coordinated by Andrea Gombar based at Los Angeles Unified School District, Los Angeles, California.

Colorado
Bookcliff Middle School, Grand Junction, Colorado, 1991–92
East Middle School, Grand Junction, Colorado, 1991–92
Fruita Middle School, Grand Junction, Colorado, 1991–92
Orchard Mesa Middle School, Grand Junction, Colorado, 1991–92
Mount Garfield Middle School, Grand Junction, Colorado, 1991–92
West Middle School, Grand Junction, Colorado, 1991–92
Coordinated by Kathleen Kain (Site Coordinator) and Rebecca Johnson (Field-Test Teacher) based at Mesa County Schools, Grand Junction, Colorado.

Michigan
Isaac E. Crary Middle School, Waterford, Michigan, 1990–92
Detroit Country Day School, Birmingham, Michigan, 1990–92
Stevens T. Mason Middle School, Waterford, Michigan, 1990–92
John D. Pierce Middle School, Waterford, Michigan, 1990–92
Coordinated by Rochelle Rubin based at the Instructional Materials Center, Waterford, Michigan, and David Housel based at Waterford Public Schools, Waterford, Michigan.

Missouri
Academy of Arts & Sciences, Kansas City, Missouri, 1990–91
Coordinated by Francesca Mollura based at the Academy of Arts & Sciences, Kansas City, Missouri.

North Carolina
Farmville Middle School, Farmville, North Carolina, 1991–92
Coordinated by Brenda Evans based at the Department of Public Instruction, Raleigh, North Carolina.

Pennsylvania
Davis School at IUP, Indiana, Pennsylvania, 1991–92
Freeport Junior High School, Freeport, Pennsylvania, 1990–92
Milton Hershey School, Hershey, Pennsylvania, 1991–92
North Hills Junior High School, Pittsburgh, Pennsylvania, 1991–92
Coordinated by Thomas Lord (Site Coordinator) and Terry Peard (Assistant) based at Indiana University of Pennsylvania, Indiana, Pennsylvania.

Wisconsin
Lundahl Junior High, Crystal Lake, Illinois, 1991–92
North Junior High, Crystal Lake, Illinois, 1991–92
Richfield Senior High, Richfield, Minnesota, 1991–92
Wilbur Wright Middle School, Milwaukee, Wisconsin, 1990–92
Coordinated by Jean Moon (Site Coordinator, 1989–90), Craig Berg (Site Coordinator, 1991–92), and Tracy Posnanski (Graduate Assistant) based at University of Wisconsin-Milwaukee, Milwaukee, Wisconsin.

Program Reviewers

Michael R. Abraham, *University of Oklahoma, Norman, Oklahoma* (Science Content, Instructional Model)
Thomas Anderson, *University of Illinois, Champaign-Urbana, Illinois* (Reading)
Albert A. Bartlett, Professor Emeritus, *University of Colorado, Boulder, Colorado* (Science Content)
Clyde R. Burnett, *Fritz Peak Observatory, Rollinsville, Colorado* (Science Content)
Elizabeth Beaver Burnett, *Fritz Peak Observatory, Rollinsville, Colorado* (Science Content)
Kallene Casias, *Turman Elementary School, Colorado Springs, Colorado* (Cooperative Learning)
Audrey Champagne, *SUNY, Albany, New York* (Instructional Model)
Aileen Dickey, *Wildflower Elementary School, Colorado Springs, Colorado* (Cooperative Learning)
Peter Drotman, *Centers for Disease Control, Chamblee, Georgia* (Science Content)
Richard A. Duschl, *University of Pittsburgh, Pittsburgh, Pennsylvania* (Nature of Science, Science Content)
Diane Ebert-May, *Northern Arizona University, Flagstaff, Arizona* (Science Content)
Timothy Falls, *Meadows Elementary School, Novi, Michigan* (Safety)
Robert J. Francis, *GM Hughes Electronics, Los Angeles, California* (Science Content)
Terry Gerbstadt, *KRDO, Channel 13, Colorado Springs, Colorado* (Science Content)
Jerald Harder, *Aeronomy Laboratory, National Oceanic and Atmospheric Administration, Boulder, Colorado* (Science Content)
Henry Heikkinen, *University of Northern Colorado, Greeley, Colorado* (Science Content)
Werner Heim, *Colorado College, Colorado Springs, Colorado* (Science Content)
Jane Heinze-Fry, *Cornell University, Ithaca, New York* (Science Content)
Sheryl Hobbs, *Carmel Middle School, Colorado Springs, Colorado* (Cooperative Learning)
Martin Hudson, *Hughes Aircraft, Denver, Colorado* (Science Content)
Jack Lochhead, *Ventures in Education, New York, New York* (Instructional Model)
James McClurg, *University of Wyoming, Laramie, Wyoming* (Science Content)
Joseph D. McInerney, *BSCS, Colorado Springs, Colorado* (Science Content)
Verjanis Peoples, *Grambling University, Grambling, Louisiana* (Equity)
E. Joseph Piel, *Professor Emeritus, SUNY, Stony Brook, New York* (Science Content)
Belinda Rossiter, *Baylor College of Medicine, Houston, Texas* (Science Content)
Kathleen Roth, *Michigan State University, East Lansing, Michigan* (Instructional Model)
Frank Tallentire, *Aerospace Engineer, Retired, Littleton, Colorado* (Science Content)
Lynn Williams, *University of Oklahoma, Norman, Oklahoma* (Nature of Science)

Artists and Photographers for Field-Test Editions

Susan Bartel
Carlye Calvin
John D. Cunningham
Michelle Dinan
Carmen Franco-Stephenson
Suzanne Guthrie
Sandy Keller
John McDowell
Staff Photographs, National Center for Atmospheric Research
Jacqueline Ott-Rogers
Ed Reshke
Nancy Smalls
Bob Trochim
Linn Winsted Trochim

Other BSCS Staff Contributing to the Project

Cindy Anderson
Debra Hannigan
Michael R. Hannigan
Sandy Keller
Joseph D. McInerney
Jean P. Milani
Dee Miller
Dee Nolan
Carolyn O'Steen
Judy Rasmussen
Bruce Thompson
Pam Thompson
Katherine A. Winternitz
M. Jean Young

Coordination, Text Design, Electronic Production and Prepress   PC&F, Inc.

Public Support   National Science Foundation

Private Support

Science Kit & Boreal Laboratories, Inc., Tonawanda, New York   IBM Educational Systems, Atlanta, Georgia

Index

A

Adams, Thomas, 195
Adhesion, 252
Adhesive forces, 251–252
Advertising, 107
Afterimage, 35–38
Agriculture, and genetic engineering, 292
AIDS, 293–294
  and DNA, 294
Airplanes, 169
Alchemy, 219
Alcohol, consumption of, and driving, 89–93
Alleles, 277–279
American Automobile Association, 181
Anaxagorus, 214
Animal senses, limits and diversity in, 21
Animation, 50–52
Aristotle, 208–209, 219
Atoms, 214
Automobiles
  design of, 181
    tires, 198
Average, 31–33

B

BAC levels, 90–92
Bacteria, 291–292
Bats, 21
Bell curve, see Curve, normal
Blind spot, 39
Blood Alcohol Content (BAC), 90–92
Blood sugar, 292
Boats, 132
  building, 145–146
  constraints for, 134
  criteria of, 134–136
  propulsion of, 136–145
  Tom Thumb, 132–133, 251–252
  types of, 133
Body cells, 271–272, 275
Body shape, 280
Bohr, Niels, 215
Bony tori, 264, 275–276
Bounceability, 198
  diversity of, 188–190
Bradbury, Ray, 299
Breathalyzer, 92

C

Camptodactyly, 264
Canoes, 135
Cells, 271–272
  body, 271, 275
  nucleus of, 271–272, 276
  sex, 271
Cereals, comparing, 125–126
Chart
  design process, 153
  flow, 147, 153–154
  ratings, 116
Chicle, 195
Chinese philosophers, 206–209, 214
Christiansen, Godfried Kirk, 159
Christiansen, Ole Kirk, 159
Chromosomes, 271, 273–279, 281–283, 295–296
Chromosome theory of inheritance, 274–283
Clone, 293
Cloning, 293
Cohesion, 252
Cohesive forces, 251–252
Color, 37–38
  seeing, 19–20
  on TV, 56–57

Color vision, 37–38
Commercials, 107
Compatibility, and color TV, 47
Cones, in the human eye, 19–20, 37–38
Constraints, 123–124
    and design, 158, 168
    and design process, 146–152
    evaluating, 126–129
    for boats, 134
    and product diversity, 168, 176
    types of, 158
Consumer Reports, 111, 118
Consumers, 107, 111, 118
    of paper towels, 111–117
Continuous motion rate, 50–54, 97
Control variables, 14, 97
Cornea, 19
Criteria
    of boats, 134–136
    and design process, 147–152
    evaluating, 123–124, 126–129, 145
    and handicaps, 181
    of products, 168
    and toys, 159
Curie, Marie, 215
Curve, bell, *see* Curve, normal
Curve, normal, 27–28, 88, 96
    graph of, 27–28, 31
    meaning of, 30–31
    and setting standards, 87–89
    value of, 31–33

D

Dalton, John, 215
Data table, 6–10
Decisions
    of boat builders, 134
    and design process, 150–152
Definitions, operational, 15, 97, 155
Democritus, 214–215
Denmark, 159
Deoxyribonucleic acid, 271
Design
    of airplanes, 169–173
    of automobiles, 181
    and constraints, 158, 168
    and criteria, 147–152
    and diversity, 161, 173
    process of, 131, 146–154
    of products, 123–124, 157, 176
    of vehicles, 181
Design engineers, 158
Designs, similarity and diversity in, 167–169
Diabetes, 292, 296
Dimples, 265
Disabilities, people with, 178–181
Diseases, inherited, 296
Disease specialists, 294
Distance
    perception, 71–73, 79–83
    reaction, 72–73, 79–83
    skidding, 72–73, 83–84
    stopping, 83–85, 97
Diversity, 5, 27–28, 31–33
    of afterimage, 37–38
    of airplanes, 169
    in animal senses, 21
    of bounceability, 188–190
    and design, 161, 173
    human, 296
    of human limits, 41
    of limits, 9
    model for, 271–283
    and speed limits, 88
    in vision, 39
DNA, 271, 292
    and AIDS, 294
    and dog tags, 295
DNA fingerprinting, 293–294
Dominance, 282
Dominant trait, 278
Dots, on TV screen, 45–47
Drinking, and driving, 90–93
Driving
    and alcohol consumption, 91–93
    and the handicapped, 181
Driving under the influence (DUI), 90–93
D/s ratio, 56–57
Dugouts, 135
DUI, 90–93
Duplo, 159
Du Pont laboratories, 191

E

Earlobes, 263, 266
Echo-location, 54–56
Eggs, female, 292–293
Electron gun, 46–47, 64
Electrons, 46
Elements, 207–209, 214, 219
Empedocles, 208–209
England, 292–293
Environment, evaluating, 177–178
Ergonomics, 157
Evaluation
 of constraints, 126–129
 of criteria, 123–124, 126–129, 145
 of environment, 177–178
 of model, 249–250
Evidence, 216–217
Explanations, changes in, 214–215
Eyeball, 19
Eyesight
 of bats, 21
 human, 19–20

F

Factors, human, 60–61, 63–65, 118, 123, 157, 177–178
 and design constraints, 158
Fahrenheit, 451, 299
Family tree, 268–270, 277, 285
FCC, 47, 65
Federal Communications Commission (FCC), 47, 65
Flicker-Fusion Frequency, 54–56, 97
Flies, 21
Flow charts, 147
 and design process, 153–154
Fovea, 20
Freckles, facial, 262, 264, 275
Frogs, 21

G

Gas, 142–145
Gender, determining, 292–293
Genes, 274–275, 291, 296
Genetic engineering, 291–297
 and society, 296–297
Genetic engineers, 291–293
Geneticists, 274, 278, 282, 293, 295–296
Genetics, future of, 295–296
Genome, 295–296
Goals, 121–122
 of products, 123
Goldfish, 21
Graph, 31–33, 36, 37
 normal curve, 27–28, 88
 of peripheral vision, 27–28
 time and distance, 77, 82
Greek philosophers, 206–209, 214
Gregoire, Marc, 191
Gum, 195

H

Handicaps, people with, 178–181
Hardness, 191–193, 199
HDTV, 65
High Definition TV (HDTV), 65
History, family, 268–271
HIV, 293–294
 and DNA, 294
Houdini water, 220, 222–223
Human factors, 60–61, 63–65, 118, 123, 157, 177
 and the handicapped, 181
Human Genome Project, 295–296

I

Ideas, testing, 216
IDTV, 65
If–then statement, 239–240
Images, visual, 19–21
Improved Definition TV (IDTV), 65
Industries, 157
Inherit, 271
Inheritance, chromosome theory of, 274–283
Insulin, 291–292
Interlacing, 64–65
Iris, 19

J

Japan, TV pictures in, 64–65
Journey to the Moon, 299

L

Lego, 159
Legoland Park, 159
Lens, of human eye, 19–20
Light, 20, 37
   and translucence, 192
Limit, 5
Limits, 14, 28, 38
   in animal senses, 21
   human, 1, 71
   inner, 17
   outer, 17
   speed, 87–89

M

Manufacturers, 118
Materials
   basic, 206–209
   forms of, 222
   insides of, 203, 206–209, 211
   properties of, 190–195, 206
Measurements, range of, 29
Men, and alcohol, 91
Microscopes, 217
Military, U.S., 295
Model
   and diversity, 271–283
   particle, 217–218, 223, 231
   scientific, 216–218, 238, 275–276
   of the universe, 217
Models, 242
   creating, 3, 238–239
   evaluating, 249–250
   improving, 222–223
   and predictions, 239–240
   review of, 253
   revising, 250–251
   testing, 233–237, 238
Motion, continuous, 50–54, 97
Movies, 52–54

N

National Television Systems Committee (NTSC), 63–65
Newton, Isaac, 215
1984, 299
Noise, 97–99
Nonscience, science vs., 241–242
Normal curve, 27–28, 96
   graph of, 27–28, 88
   meaning of, 30–31
   and setting standards, 87–89
   value of, 31–33
Normality, 26
NTSC, 63–65
Nucleus, cell, 271–272, 276

O

Operational definition, 15, 97, 155
Orwell, George, 299

P

Packaging, 112
Paper towels, 111–118
Particle model, 217–218, 223, 231
Particles, 214–216, 230
   movement of, 222–223
Particle theory, 230
Perception distance, 71–73
   determining, 79–83
Perception time, 71–73
Peripheral vision, 16–17, 20, 96
   degrees of, 28
   graph of, 27–28
   and human eye, 19–20
Persistence of vision, 48–49
Philosophers, 206–209, 214
Philosopher's stone, 219
Philosophy, 219
Phosphor, 46–47
Pictures, TV, 45–47, 56–57, 64–66
Pixels, 46–47, 62
Plunkett, Roy, 191
Poly(styrene-butadiene) co-polymer, 198
Polyvinyl acetate, 195
Predictions, making, 239–240

Presentations, 147
Product design
　and constraints, 123, 168, 176
Product designers, 176
Products
　criteria of, 168
　designing, 123–124, 131, 157
　function of, 168
　goals of, 123
　judging, 118, 123
　of modern technology, 146–147
Propellers, 139–142
Properties, 187
　exploring, 196–197
　of materials, 190, 206
　review of, 253
Propulsion, 136–145

R

Range, 17, 31–33
　of measurements, 29
Ratings chart, 116
Reaction distance, 72–73
　determining, 79–83
Reaction time, 72–73, 88
　personal, 74–78
Recessiveness, 282
Recessive trait, 278
Retina, 19–21, 37–39
Rods, in the human eye, 19–20
Rutherford, E. J., 215

S

Sailboats, 135
Sails, 137–139
Santa Anna, Antonio Lopez de, 195
Science, vs. nonscience, 241–242
Science fiction, 299
Scientific models, 216–218, 275–278
Scientists, 183, 211, 215, 218–219, 230, 239, 274, 287, 294
Screens, TV, 64–65
Senses, animal, limits and diversity in, 21
Sex cells, 271, 274, 281, *see also* Eggs, female; Sperm

Shadow mask, 47
Skidding distance, 72–73, 83–84
Skidding time, 72–73
Snakes, 21
Society, and genetic engineering, 296–297
Sopadilla tree, 195
Speed
　and distance, 73–75
　limits on, 87–89
Sperm, 271, 274
Standard, 62
　and normal curve, 87–89
　setting, 87–89, 97
　of TV pictures, 64–65
Stickiness, 199
Stopping, phases of, 71–74, 97
Stopping distance, 83–85, 97
Success, measuring, 13–14
Surface tension, 251–252

T

Technology, 153, 176–177, 180
　and boats, 135
　and design process, 131, 146–154
　and genetic engineering, 292
　and human factors, 63–65, 118, 123, 157, 177, 181
　and problem solving, 146–147
　and product diversity, 168, 176
Teflon, 191
Television. *See* TV
Testability, 241–242
Tests, of models, 233–238
Tetrafluoroethylene, 191
Thales, 208
Theory, 216–217
　particle, 230
　scientific, 218
Thompson, J. J., 215
Thumbs, 265
Time
　perception, 71–73
　reaction, 72–73, 88
　reaction, personal, 72–78
　skidding, 72–73
Tires, automobile, 198

Toys, 154–157, 159
Traits, 262–267
    changing, 291
    and family history, 268–271
Translucence, 191–192, 199, 239–240
Transportation, and boats, 135
TV, 42–44, 97
    color, 56–57
    design of, 61–64
    pictures on, 45–47, 56–57
    screen of, 45–47, 64–65
    and viewing distance, 59–60
Twenty Thousand Leagues Under the Sea, 299

U

United States, TV pictures in, 64–65
United States military, 295
Universe, model of, 217

V

Variables, 14
    controlling, 14–15, 35, 97–98
    identifying, 15
Verne, Jules, 299
Viewing distance, for TV, 59–60
Viscosity, 192, 194, 199, 240
Vision, persistence of, 48–49
Vision, peripheral
    color, 37–38
    degrees of, 28
    graph of, 16–17, 20, 27–28, 96
    and the human eye, 19–20

W

Wheel of Traits, 262–267
Women, and alcohol, 91
Wrigley, William, Jr., 195